TRICKY SWITCHIN' MADE SIMPLE

This fully illustrated book details the conventional and unconventional use of multi-pole, 3-way and 4-way switches to control lighting and power.

Written in concise easy to understand text with detailed wiring diagrams. This book takes the trick out of complex switching and makes it a real treat for both you and your customer.

A must read for the experienced electrician. Shows the in's and out's of using multi-pole, 3-way and 4-way switches in both conventional wiring and also in an unconventional manner not usually taught in the trade. Do it yourselfers should also enjoy reading and applying these techniques in their own homes.

This book is written and published by Bill Eilbacher, a retired member of the International Brotherhood of Electrical Workers (IBEW), Local 617, San Mateo, California. All ideas and concepts found in this book are the sole intellectual property of Bill Eilbacher. During the time that I conceived and perfected the techniques found in this book, I also ran a successful small signatory electrical contracting firm located in Fremont, California. There I employed members of Local 595, Dublin, California up until retiring in 2008. I would now like to share with you these unusual switching techniques which I developed, by publishing and making this very informational book on tricky switching scenarios available to all.

DEDICATION

I would like to take this time to mention two specific people who were instrumental in giving me the opportunity to not only become a successful electrician but to carry that on one step further by becoming a successful union signatory electrical contractor serving Alameda County, California.

My sincerest thank you goes out to Norm Hertlein and his late brother Henry, founders of the former Hertlein Electric, Incorporated where I was employed for over twelve years.

These two wonderful gentlemen gave me the opportunity to become a foreman for their firm, in spite of my limited knowledge and immaturity. They saw something in me that even I didn't know existed. In the coming years they further nurtured it by allowing me to manage larger and more complicated projects. During my final years with the firm I was promoted to the position of Residential Contract Manager, where I oversaw all aspects of the residential side of the business. It was because of their graciousness and involvement that I eventually left the firm to start my own successful electrical contracting business. Thanks again Norm and Henry; I couldn't have done it without your encouragement and help.

Additionally, I would also like to thank the many members of Local 595, Dublin, California, that I had the pleasure of working with throughout the years, both while as a foreman for Hertlein Electric and as the owner of Eilbacher Electric. I would like to give special thanks to Jim Martin, a long time dedicated employee with whom I formed a special bond. Thanks a lot Jimbo.

INTRODUCTION

As an IBEW ®[1]electrician for well over thirty-five years, I have worked in virtually all aspects of the trade. After starting out as a maintenance electrician in a steel mill back in the early 70's, I progressed up and through the ranks as a residential, commercial and industrial inside wireman. My first twelve years as an inside wireman were primarily spent on residential wiring and troubleshooting.

During my career as an electrician, I have certainly seen my share of problematic wiring. Although the problems were diverse, many revolved around switching scenarios, especially 3-way and 4-way switching. I was constantly bewildered at the number of times I encountered three and four way switches that were either installed incorrectly or simply would not work at all. I discovered that most do it yourselfers did not have a clue when it came to wiring difficult switching arrangements. Many would often make things next to impossible for the person trying to repair the work correctly and few were installed in a code complying manner.

Therefore, I decided to prepare this book for the professional electrician, as well as the do it yourselfers having desire for a firm grasp of understanding the concepts of three and four way switching. This book does not stop there. It goes on to explain the unconventional use of multi-pole, 3-way and 4-way switches used to control lighting and power in a manner that will amaze you. You will learn how to do complicated lighting control in a simple and inexpensive manner not currently taught in the trade. In addition, this book will show you how to construct a very simple and economical ac "or" gate which will allow the user to derive 120 volt power from a source not normally nor easily converted to such use.

Chapter three is a special section explaining how even more extremely tricking switching scenarios can be accomplished by combining 3-way and 4-way switches in a very unusual way. Many professional electricians would look extremely puzzled if confronted with the manner in which theses switches were used. These tricky switching scenarios however, are easily understood and carried out in the electrical schematics on the pages detailed in chapter 3.

I hope you'll have fun and enjoy reading and applying these techniques in your profession, sideline or hobby. You will surely impress yourself, your friends, your colleagues and customers alike with the trick wiring techniques shown on the following pages. As always, make sure that all work is performed in a code complying and professional workmanlike manner. As the author and publisher of this book, I have made every effort to insure that all material presented is clear and accurate. Therefore, I shall not be held responsible for any errors in content; nor shall I be responsible for the interpretation or application of any material presented in this book. As local electrical codes may vary, always check with the authority having jurisdiction before proceeding with any electrical work.

---

1   IBEW is a registered trademark of the International Brotherhood of Electrical Workers

FORWARD

Although there are many ways to accomplish complicated or tricky switching, this book focuses on the inexpensive and unconventional use of 2-pole, 3-way and 4-way switches. In a simple easy to understand manner, this book will teach many tricks and secrets not taught in the trade, nor readily available to the skilled electrical tradesperson. Although some of these same techniques can be accomplished by the use of relatively expensive hardware using proprietary methods and materials, this book will concentrate on the use of simple, economically feasible and readily available switches. These easy to find switches, when used with hardwiring techniques, will assure that your system will operate for years to come. When a high cost proprietary system fails and components need to be replaced, they may no longer be available to repair the work. Due to new technology, functional obsolescence or limited acceptance, the replacement parts needed may not be available.

With a hardwired system however, that's not a problem. Simple switches such as 2-pole, 3-way and 4-way have been around for ages and will continue to be around for the foreseeable future. So why not put your trust into something that you can understand, rather than technology that you'll need an electrical engineering degree to master.

This book will thoroughly explain the basics of multi-pole, 3-way and 4-way switching and then go on to the real fun of using those commonly available switching devices in a foolproof manner that will make complicated and tricky switching easy. For those already thoroughly familiar with and who understand the basics of multi-pole, 3-way and 4-way switches, simply skip to the advanced sections of this book.

Chapter three in the advanced sections of this book will explore even more difficult and extremely tricky switching scenarios. Drawings and text detailing the special combinations of using 3-way and 4-way switches together in a very unconventional manner should amaze the reader. It will allow the reader to perform extremely complex switching tasks in a manner not generally understood by even the most experienced electricians.

TABLE OF CONTENTS

CHAPTER 1: BASICS 1
Train yard analogy 2
Understanding travelers 4
Understanding 3-way switching techniques 6
Deadending 3-way switches 8
Maintained contact single-pole double-throw 12
Maintained contact center position off 14
Momentary contact center position off 16
Understanding 4-way switching techniques 18
Understanding multi-pole switches 20

CHAPTER 2: TRICKY SWITCHING
Override selected exterior lighting from one location 24
Override selected exterior lighting from two locations 28
Override selected exterior lighting from three or more locations 30
Override common exterior lighting 32
Control selected night lighting from one location 36
Control selected lighting from two locations with night lighting from one location 38
Control selected lighting from three or more locations with night lighting from one location 40
Control selected lighting from three or more locations while controlling three night lights
from their own individual locations 42
3-way switch used as a simple transfer switch 46
Knob and tube style three way switching 48
Existing out-building lighting control and deriving power from existing three
way circuit by using an ac "or" gate 50
4-way switch used as dc motor reversing switch 52
Custom kitchen lighting control using 2-pole switches 54

CHAPTER 3: EXTREMELY TRICKY SWITCHIN'
Introduction to combining 3-way and 4-way switches 58
Extreme override control used with multiple common exterior lighting scenarios 60
Extremely tricky control of night lighting from two or more locations 66
Extremely tricky control of multiple night lighting from two or more locations 70
Extremely tricky control of custom kitchen night or task lighting from two or more locations 74

CHAPTER 4: FORMULAS AND OTHER USEFUL INFORMATION
The magic triangle, Ohms Law and Power formulas 78
STOP!…Do not open that neutral 80
Wire twisting, is it really necessary? 82
Easiest way to strip non-metallic sheathed cable 83
Resistor Color Code 84
California three way switching 86
Well pressure tank controls 88
Other proprietary systems 90
A brief explanation of X-10 technology 90
Common electrical symbols and diagrams 91

GLOSSARY
Terms 93-100

electrician

Chapter 1
BASICS

TRAIN YARD ANALOGY

UNDERSTANDING TRAVELERS

UNDERSTANDING 3-WAY SWITCHING TECHNIQUES

DEAD ENDING 3-WAY SWITCHES

MAINTAINED CONTACT SIGLE-POLE DOUBLE-THROW

MAINTAINED CONTACT CENTER POSITION OFF

MOMENTARY CONTACT CENTER POSITION OFF

UNDERSTANDING 4-WAY SWITCHING TECHNIQUES

UNDERSTANDING MULTI-POLE SWITCHES

TRAIN YARD ANALOGY

When comparing a 3-way switch to a train switching device, it's easily understood. Picture a railroad yard where one track comes in and one track goes out as depicted in figure A-1 on the following page. In the middle of the yard is an area used for maintenance work performed on the trains coming into the yard. Some trains need to continue on their path while others need to be brought into the yard for maintenance. Depending on the position of the track switchers, the trains can either go straight through and bypass the yard (A position) or be detoured to the maintenance area (B position).

While in the maintenance area, other trains can go through once the track is switched back to the A position. After the maintenance is completed, the switch at the other end of the yard is changed back to the A position. This will redirect the train and allow it to continue on its path away from the maintenance area. In this manner of thinking, the two sets of tracks can be compared to the travelers in the 3-way circuit.

A 3-way switch is basically the same. As shown in figure A-2, the switch takes the power coming in and redirects it onto one of the travelers going out. In one position, the power, just like the train, would come into the switch and continue on its way out on one traveler while in the other position (figure A-3), the power, like the train, would be redirected onto the other traveler where it would sit idle, waiting for the other end to be switched. Once the other end of the 3-way switch is toggled, the power continues on its way.

"A" POSITION
ALLOWS TRAINS TO GO THROUGH YARD

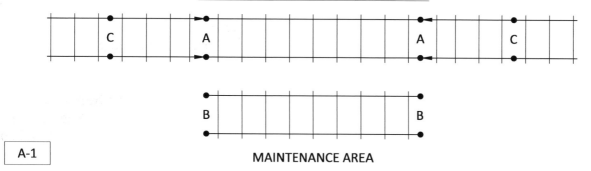

MAINTENANCE AREA

A-1

- OR -

"B" POSITION
DETOURS TRAINS INTO MAINTENANCE AREA

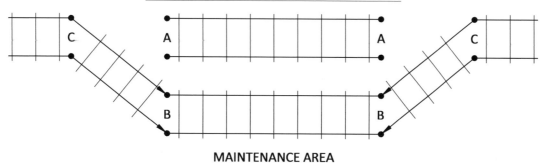

MAINTENANCE AREA

A-2

A-3

UNDERSTANDING "TRAVELERS"

Most people think of travelers as people going from one place to another. In 3-way and 4-way switching, the "travelers" are the two wires which allow the electrical circuit to go from one place to another. The two travelers are connected to the two terminals which we'll call A and B on each 3-way switch. These two terminals are not the common terminal. These two terminals are generally lighter in color while the common terminal is usually darker in color and sometimes marked with the word common or letter C. The job of the travelers is to be either hot or not. While one traveler is hot, the other is not. Toggling the 3-way switch to the opposite position will change the state of the travelers. That means that the traveler that was hot before is now not and the traveler that was not hot before now is.

In most modern residential wiring systems, the travelers are grouped together with a third wire in a single 3-conductor jacketed cable. Depending on the application, this third conductor may be used as either a hot wire or a neutral wire. The third conductor is always colored white while the two traveler conductors are always colored black and red. Some electricians mistakenly use the red and white combination or the black and white combination as the travelers. This practice however, is incorrect and is also considered an electrical code violation. This third white wire in the cable assembly if used as a hot conductor rather than a neutral conductor is usually taped black or some color other than green. This coloring or marking must be located at all exposed ends where the wire leaves the cable jacket. This colored marking readily identifies the white wire as something other than a neutral conductor and in addition, the most recent edition of the National Electrical Code ®[2] (NEC) requires that this be done in the interest of safety. In commercial applications the travelers can be any color other than white or green. In many cases the travelers will both be the same color, usually a different color than the hot power conductor.

For ease in troubleshooting and also simply for aesthetics, always swap the travelers at the opposite ends of the two separate 3-way switches. For clarity, assume that each 3-way switch has three terminals marked A, B and C. As indicated before, the C terminal is the common terminal and therefore it's internal switching contact is common to terminals A and B. With the switch in the up position as shown in B-1 and B-2, the common terminal C is connected through the switch to terminal B. Reversing the switch to the down position redirects the common terminal C to terminal A. Swapping the travelers at each three way switch means that on the first switch we would connect the black wire to terminal A and the red wire to terminal B, while on the second switch we would reverse that order and connect the red wire to terminal A and the black wire to terminal B. In this manner, illustration B-3 and B-4 will show that when both switches are in the same position the light is off. When both switches are in the down position the light is off. When both switches are in the up position the light is also off. Anytime the switches are in opposite directions, one up and one down, the light is on. This simplifies any needed troubleshooting as having to try and figure out if the light is supposed to be on or off is eliminated. This procedure also looks better and shows a sign of professionalism. It's similar to having all your trim screws aligned in the same direction as it stands out and takes notice.

---

2   National Electrical Code (NEC) is a registered trademark of NFPA

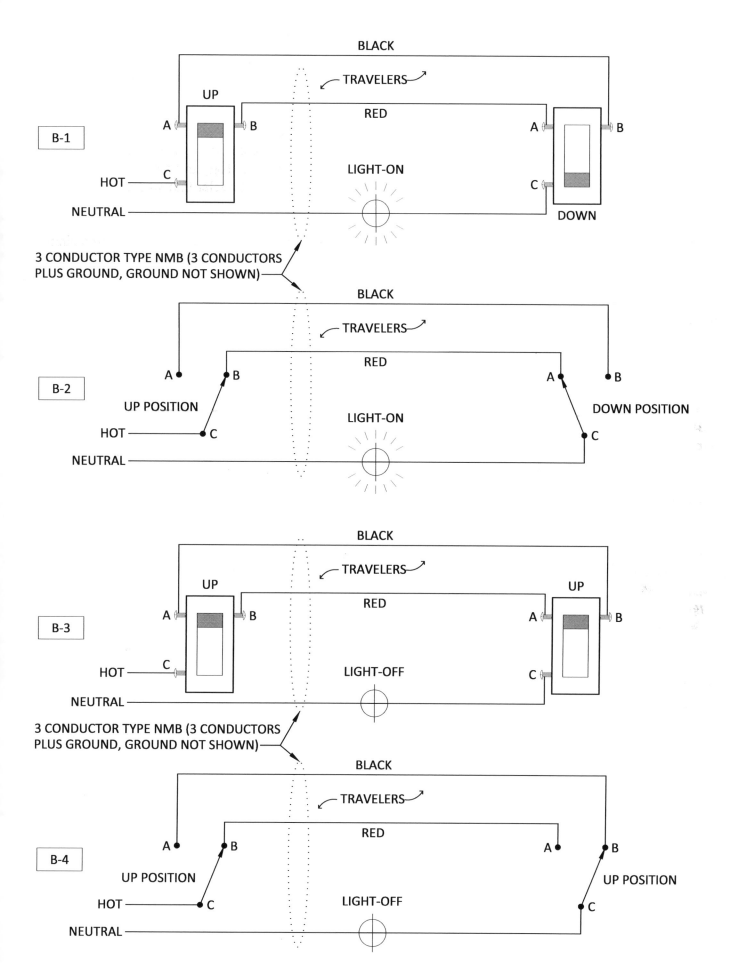

BLACK

TRAVELERS

B-1

UP

A   B

RED

HOT   C

LIGHT-ON

NEUTRAL

A   B

C

DOWN

3 CONDUCTOR TYPE NMB (3 CONDUCTORS
PLUS GROUND, GROUND NOT SHOWN)

BLACK

TRAVELERS

B-2

A   B

RED

UP POSITION

HOT   C

LIGHT-ON

A

C

DOWN POSITION

NEUTRAL

BLACK

TRAVELERS

B-3

UP

A   B

RED

HOT   C

LIGHT-OFF

NEUTRAL

UP

A   B

C

3 CONDUCTOR TYPE NMB (3 CONDUCTORS
PLUS GROUND, GROUND NOT SHOWN)

BLACK

TRAVELERS

B-4

A   B

RED

UP POSITION

HOT   C

LIGHT-OFF

A   B

UP POSITION

C

NEUTRAL

5

UNDERSTANDING 3-WAY SWITCHING TECHNIQUES

The name 3-way switch is a misnomer. A 3-way switch by itself can only control lighting from two different locations. In order to control lighting from three or more locations, 4-way switches must be incorporated into the switching circuit. A 3-way switch is simply a single pole double throw wiring device. Single pole indicates that it can only control one conductor while double throw indicates that the controlled conductor can be switched from a common terminal to either one of the other two terminals. One terminal is open while the other terminal is closed. Change the position of the switch and the open circuit and closed circuit change positions. A 3-way switch used in standard residential and commercial wiring has no off position. Looking at the switch, it's apparent that it does not say on or off like a standard single pole wall switch does. That's because it's never off. Just because the toggle is in the down position doesn't mean that the switch is off. It indicates that the switch is in a different position and that the state of the switch has changed. The power on the travelers have changed. As explained in "UNDERSTANDING TRAVELERS", the 3-way switch has three connections on it. One common C terminal and two traveler terminals which are referred to as terminals A and B. In some rare instances there may be a 3-way switch with four terminals on it but in those cases, the extra terminal is simply another C or common terminal located on the other side of the switch for ease in wiring.

In the diagrams on the next page, simple 3-way switch setups consisting of (2) 3-way switches, one at either end of a hallway are shown. Both of these switches control the light fixture in the center of the hallway. In diagram 1-A, the power enters the switch box on the left while the light fixture is connected to the switch box on the right. In between the two switch boxes is the cable containing the travelers. The cable between the two switch boxes is usually a 14/3, type NMB, 3-conductor with ground wire. For simplicity and clarity the bare ground wire is not shown in any of the connection diagrams. Both the cables feeding the power to the first switch box and the switch leg to the light fixture box are 14/2, type NMB, 2-conductor with ground wire.

Notice that the power coming in via the 2-conductor cable has the black wire connected to the C or common terminal while the white wire is spliced to the other white wire contained in the 3-conductor cable going to the other switch. This splice carries the white neutral wire over to the other switch box where it is also spliced to the white wire on the 2-conductor cable going to the light fixture. The black wire on the cable leading to the light fixture is then connected to the C or common terminal on the other 3-way switch. The two other wires in the 3-conductor cable are attached to the A and B terminals on each three way switch. These two other wires are the *travelers* and are always colored black and red in a 3-conductor type NMB cable.

Referring to the "UNDERSTANDING TRAVELERS" section it is now easy to see the role of the travelers in the 3-way switching circuit. In diagram 1-A, when the first switch is in the down position, the hot terminal C is connected through the switch to terminal A. From there the power flows through the black traveler where it is connected to terminal B on the other switch. Remember that the travelers are always swapped on each switch. With the second switch also in the down position, no connection is made from terminal B through the switch to terminal C which is connected to the light fixture. In this condition the light fixture is off as there is no power present on the red traveler. Toggle either switch to the up position and the light fixture illuminates. If the first switch is changed to the up position as illustrated in diagram 1-B, the hot terminal C is now connected through the switch to terminal B. The power now flows through the red traveler to the other switch where it is connected on terminal A. With the second switch still in the down position, terminal A is now connected through the switch to terminal C and the light fixture illuminates. If the first switch remains in the down position while the second switch is toggled to the up position as shown in diagram 1-C, the power flowing through the black traveler that is connected to terminal B on the second switch would now be redirected to terminal C, causing the light fixture to illuminate. As these diagrams show, the importance of wiring the travelers correctly in the 3-way switching circuit is paramount. Connecting the travelers to anything other than terminals A and B will cause erratic and faulty operation.

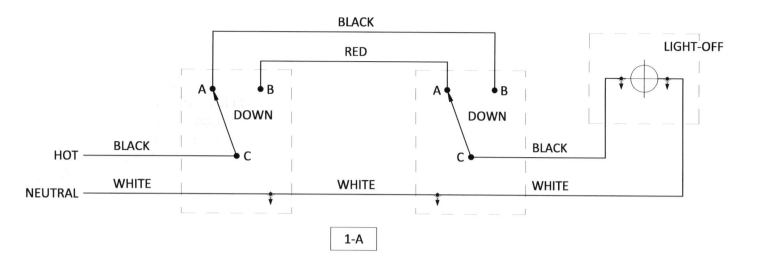

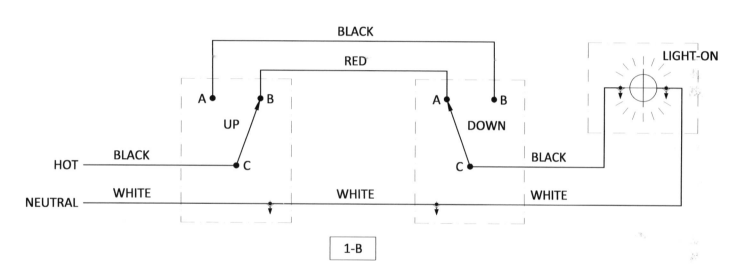

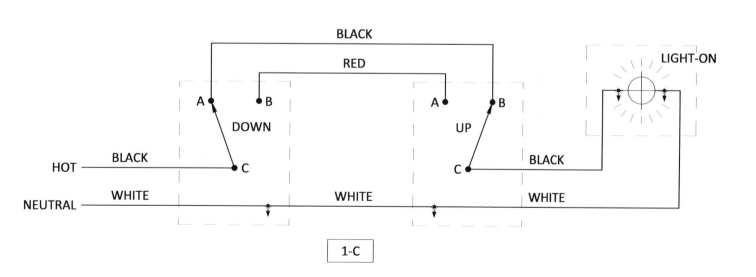

DEAD ENDING 3-WAY SWITCHES

There are many instances in which 3-way switches are dead ended. Dead ended does not mean that the power is dead. It simply means that there are no other wires leaving the switch box other than the three wires connected directly to terminals A, B and C on the 3-way switch. Usually, dead ending is done for convenience such as at the bottom or top of a set of stairs. When there is one box located at the top and another at the bottom that is used only for switching the light controlled by the 3-way switches, dead ending makes sense. There are basically two ways to dead end a 3-way switch. The first is when both the power supplying the switch and the switch leg going to the light fixture originate in the same box as depicted in figure 2-A. The second way is when the light outlet box is located between the 3-way switch boxes as in figure 2-B.

As shown in figure 2-A, the power coming in on the black wire is spliced to the white wire in the 3-conductor cable containing the two travelers. This cable then travels between the two switch boxes. Now the white wire is the hot wire and must be re-identified with some colored tape other than green. Then it's connected to the common terminal of the 3-way switch located at the other end of the hallway. The hot wire will now be redirected from one traveler to the other via the 3-way switch. Depending on the position of the 3-way switch, one traveler will be hot while the other is not. Changing the position of the switch will also change the state of the travelers. Back at the originating end of the 3-way switch loop, the black wire going to the light fixture is connected to the common terminal of the 3-way switch. As observed in diagram 2-A, the far end of the three way switch is dead ended as it has no other wires entering or leaving the box other than the three wire cable originating from the first switch box.

Another popular way of dead ending 3-way switches is shown in figure 2-B. In this scenario the light outlet box is located between the (2) 3-way switches. This arrangement might be used in a long hallway having a light fixture in the center. Observe how the 3-conductor cable is run through the light box with the black and red travelers being spliced black to black and red to red. The white wires are then attached to the light fixture with one white wire being the neutral conductor while the other white wire now becomes the switch leg. A switch leg is the conductor supplying power to the light fixture from the switch. Again in this case, the white switch leg conductor must be re-identified with colored tape to show that it is now a hot conductor rather than a neutral conductor. This procedure is required at all locations where the jacket of the cable has been removed. On the close end of the 3-way switch, the hot wire coming in is not spliced to the white wire in the 3-conductor cable as it was in the first instance as shown in figure 2-A. Rather, it is hooked directly to the common terminal of the 3-way switch while the white wire is now spliced to the white wire coming in. Subsequently, the white wire in the 3-conductor cable now enters the light fixture box where it is directly connected to the white wire on the light fixture. The other two wires, black and red, are spliced color to color and proceed to the far end of the 3-way switch loop where they are connected to terminals A and B on the 3-way switch. The white wire now re-colored with tape and referred to as the switch leg is connected to the common terminal on the far side 3-way switch. The re-identified white wire now supplies power to the light fixture when connected to the hot traveler through the 3-way switch.

It is extremely important to remember that the second arrangement as shown in figure 2-B should never be used when more than one light fixture is located between the 3-way switches. This faulty scenario as shown in figure 2-C would connect the two light fixtures in series. The voltage supplying the fixtures would then be split between the two light bulbs making them dim due to partial voltage being placed across them. If the circuit is to be extended to another light fixture location, simply install another 2-conductor cable from one light fixture box to the other as shown in figure 2-D.

As depicted in the drawings, dead ending 3-way switches is easy and saves time and money. A 3-way switch can also be dead ended in a box containing other switches or wiring devices. A common example of this technique would be at a stair landing or hallway change as shown in figures 2-E and 2-F.

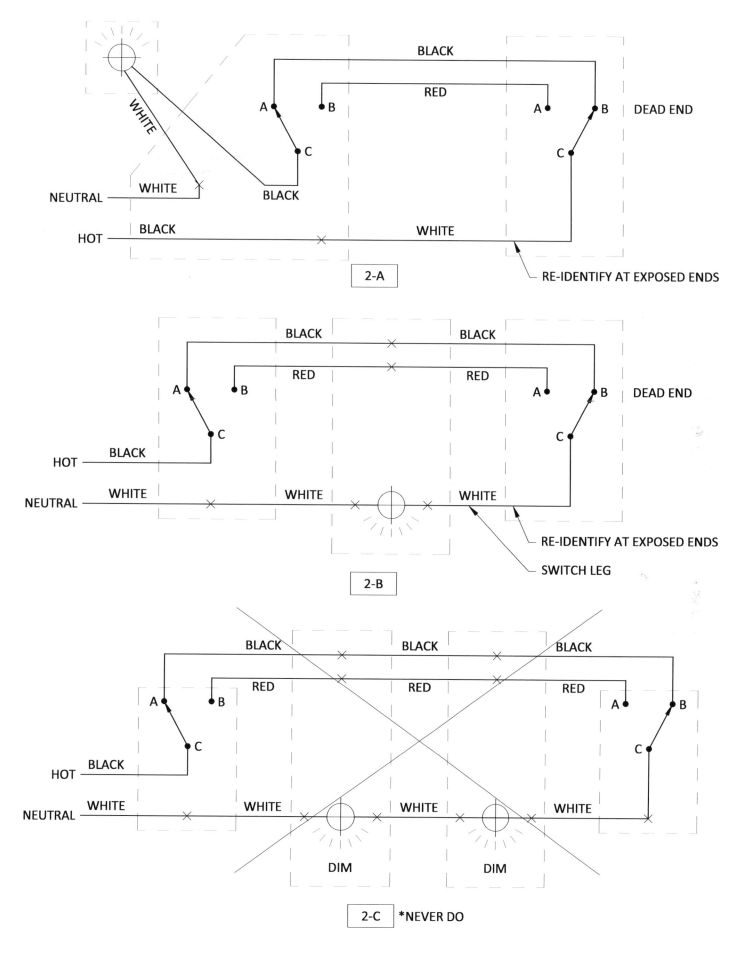

BLACK

RED

A B   A B DEAD END

C   C

WHITE

NEUTRAL   WHITE ✕

HOT   BLACK   ✕   WHITE

2-A

RE-IDENTIFY AT EXPOSED ENDS

BLACK   ✕   BLACK

RED   ✕   RED

A B   A B DEAD END

C   C

HOT   BLACK

NEUTRAL   WHITE ✕   WHITE   ✕   ✕   WHITE

2-B

RE-IDENTIFY AT EXPOSED ENDS

SWITCH LEG

BLACK   ✕   BLACK   ✕   BLACK

RED   ✕   RED   ✕   RED

A B   A B

C   C

HOT   BLACK

NEUTRAL   WHITE ✕   WHITE   ✕   ✕   WHITE   ✕   ✕   WHITE   ✕

DIM   DIM

2-C   *NEVER DO

9

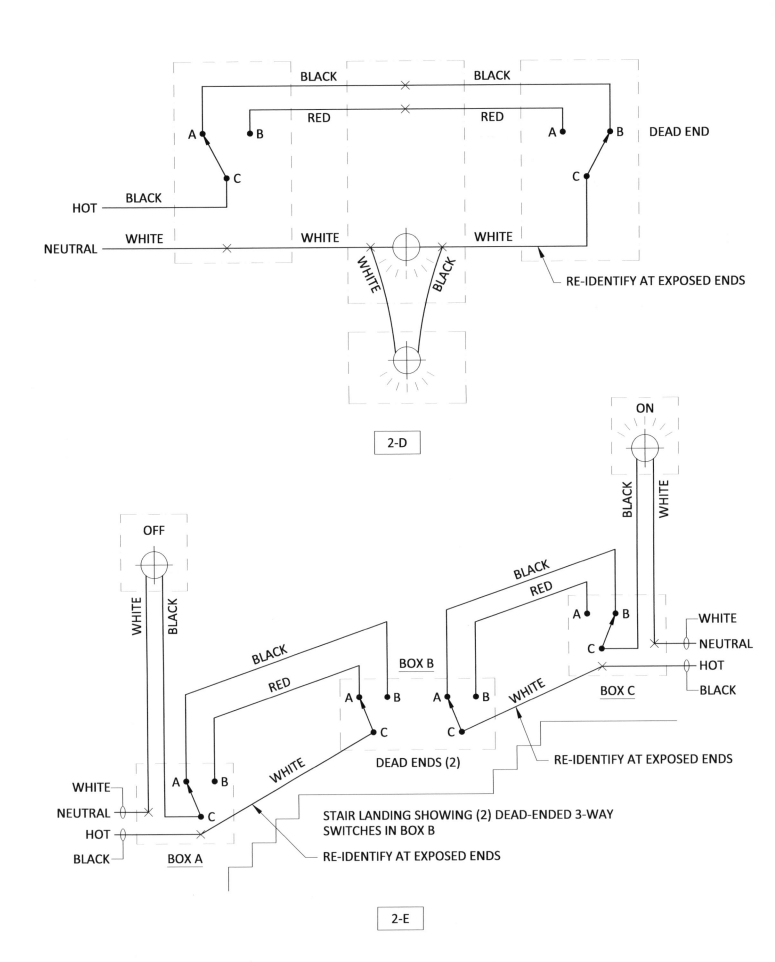

2-D

2-E

10

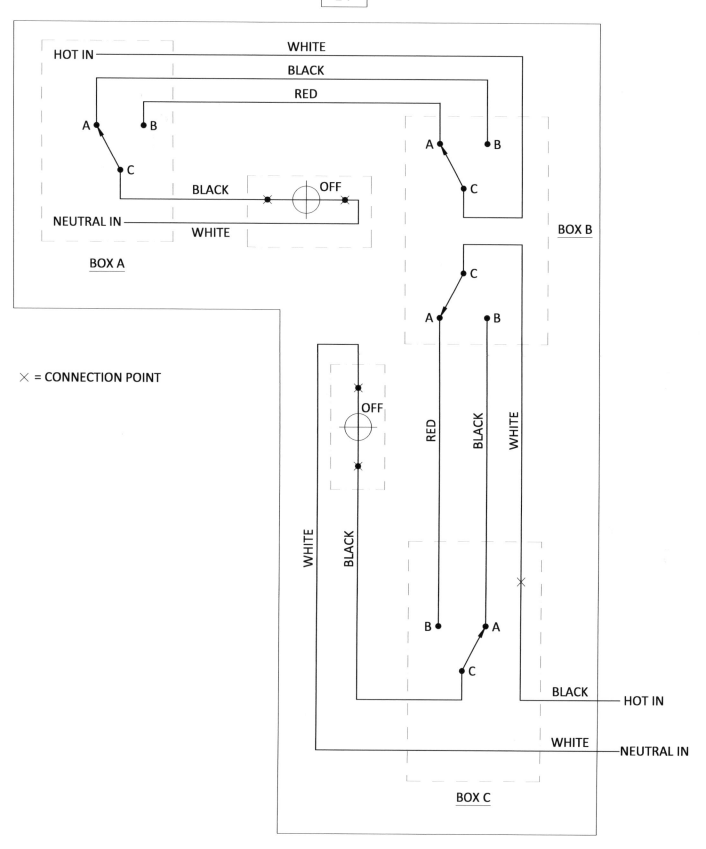

× = CONNECTION POINT

MAINTAINED CONTACT SINGLE POLE -DOUBLE THROW

There are several styles of 3-way switches with the most common type being the maintained contact single pole-double throw. This readily available switch is used in all residential and commercial applications to control lighting from two different locations. As stated before, it has no off position. Figure 3-A illustrates how the switch has two positions, up or down. In the first position with the toggle up, the common terminal C is connected internally through the switch contacts to terminal B. When the position of the toggle is changed to down, terminal C is now connected to terminal A. In conventional 3-way switch wiring, terminals A and B on the switch are always connected to the travelers while the common terminal C is connected to either the hot wire feeding the switch or to the switch leg feeding the light fixture.

In the unconventional employment of 3-way switches, this particular type of switch has several other potential uses. For example, it can be configured as a simple transfer switch utilizing two sources of power or reverse its normal use, thereby controlling two or more different light fixtures from one source. In addition to performing these two unconventional tasks, other ways to utilize these simple switches to accomplish complex switching tasks will be discussed later on in this book.

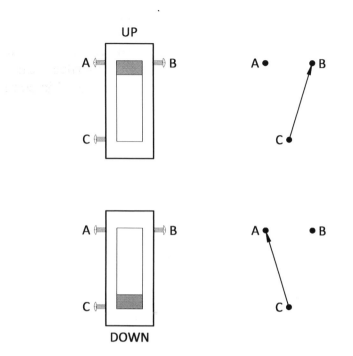

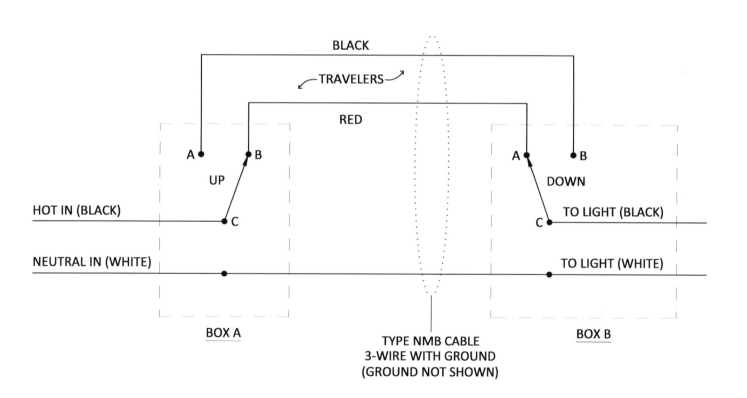

BLACK

TRAVELERS

RED

UP

DOWN

HOT IN (BLACK)

NEUTRAL IN (WHITE)

TO LIGHT (BLACK)

TO LIGHT (WHITE)

BOX A

BOX B

TYPE NMB CABLE
3-WIRE WITH GROUND
(GROUND NOT SHOWN)

MAINTAINED CONTACT -CENTER POSITION OFF

A maintained contact-center position off switch as shown in figure 4-B is identical to a standard 3-way switch except that it can be turned off. It has three positions; up, down and center off. The off position makes this type of switch an ideal candidate for specialty uses.

One such use might be in the control of several different lighting fixtures from one source of power such as in a dark room used for photography as pictured in diagram 4-A. On the inside of the darkroom there may be a white light source used for general illumination and a red light source used for film development. On the outside of the darkroom might be an illuminated "DO NOT ENTER" sign used to warn others that they should not open the door as film is being processed.

Figure 4-C indicates how this style of maintained contact-center position off switch is used for this application. The incoming power connects to the common terminal of the switch. The white light fixture connects to terminal A while the red light fixture along with the exterior "DO NOT ENTER" sign connects to terminal B. In this manner, with the switch in the down position* (C to A), the white light inside the dark room would be illuminated while the red interior light and exterior warning light are both off. If the switch is then toggled to the up position* (C to B), the white light goes off while the interior red light and exterior "DO NOT ENTER" warning sign illuminate. With the toggle in the center off position, all light fixtures located inside and outside the room are off.

* When switch diagrams in this book are drawn sideways or horizontally, the position of the arrow is not indicative of the position of the switch. All 3-way switches referred to and drawn in this book are considered to be in the down position when the common terminal C is connected to terminal A. Whenever the common terminal C is connected to terminal B, the switch is considered to be in the up position. Since a 3-way switch neither says on nor off, the switch can simply be removed and inverted 180 degrees to change the function.

MAINTAINED CONTACT - CENTER OFF (TYPICAL DARKROOM)

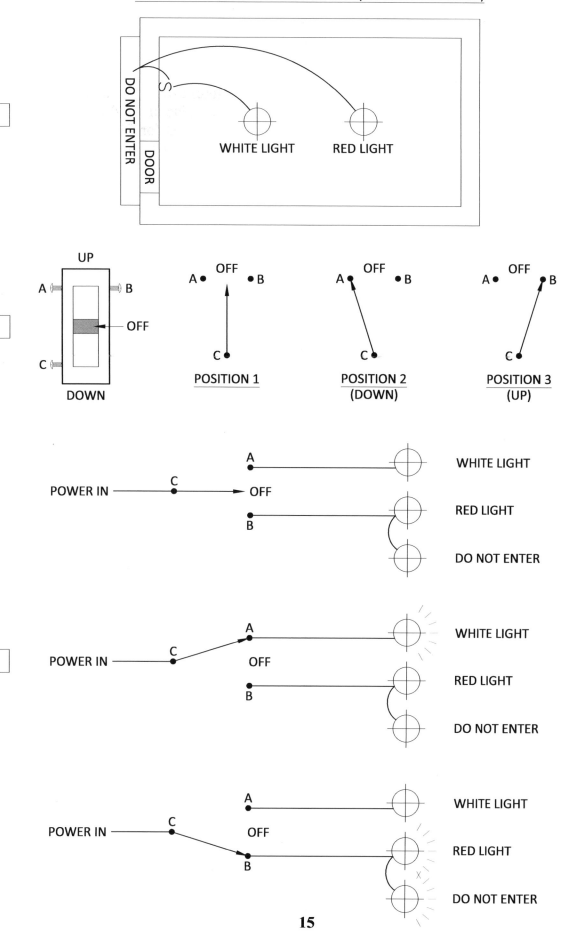

4-A

4-B

4-C

15

MOMENTARY CONTACT-CENTER POSITION OFF

    Momentary contact-center position off style is another type of 3-way switch. This particular switch is identical to the maintained contact-center position off style except that the toggle is spring loaded and only makes contact as long as it's held in either the up or down positions. Releasing the switch breaks the contact and returns the toggle to the center position which is off. This style switch is generally used to control low voltage relay controlled lighting. In this type of lighting, the power for the fixtures is usually 120, 277 or 480 volts while the voltage used to control the lighting is only 24 volts.

    The relays used to control this lighting have a high voltage 2-wire connection that acts as the switch. The low voltage side requires three wires which control the state of the relay. One wire is the common and the other two wires are used to toggle the relay on or off. As shown in figure 5-A, one wire from the low voltage 24 volt transformer is connected to the common wire on the relay. The other transformer wire is attached to the C or common terminal of the momentary contact-center position off style switch. The two other terminals on the momentary switch are connected to the two other wires on the relay. Up position on the switch turns the relay on while down position turns the relay off.

    This form of wiring allows for the use of low cost bell wire on the low voltage side of the relay and also for the momentary switch wiring. This manner of switching does not necessitate the use of travelers as in normal 3-way and 4-way switch loops. For lighting control desired at more than one location, simply run another set of low voltage wires from another momentary style switch to the same relay. Parallel wiring between the switches may also be used to accomplish this same task and can be configured for multiple locations. Diagram 5-B shows a typical arrangement employed for controlling lighting at multiple locations utilizing momentary contact-center position off 3-way switches.

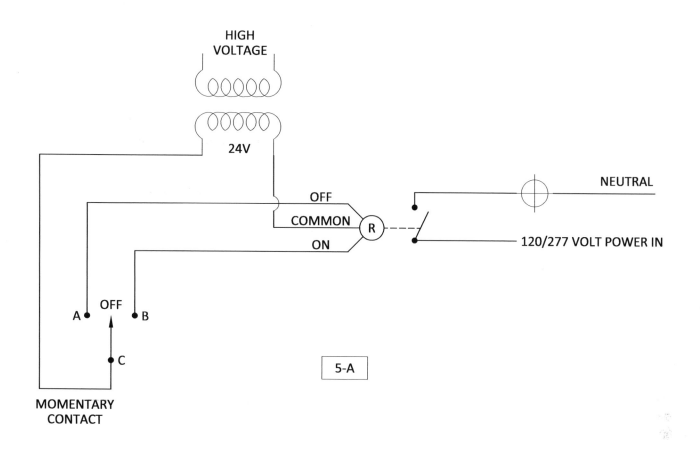

HIGH VOLTAGE

24V

OFF

COMMON

R

ON

NEUTRAL

120/277 VOLT POWER IN

A

OFF

B

C

MOMENTARY CONTACT

5-A

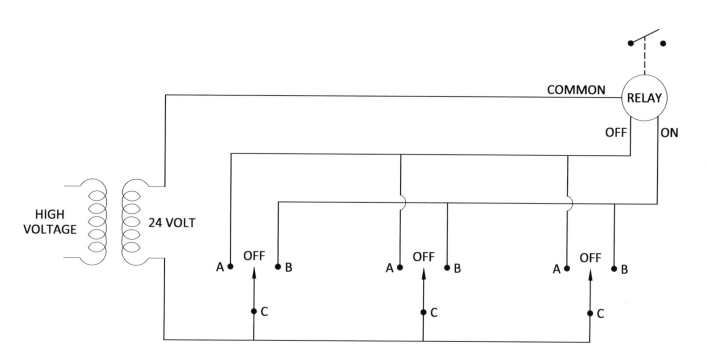

HIGH VOLTAGE

24 VOLT

COMMON

RELAY

OFF

ON

A OFF B

C

A OFF B

C

A OFF B

C

MOMENTARY CONTACT - CENTER OFF 3 POSITION SWITCHES

5-B

UNDERSTANDING 4-WAY SWITCHING TECHNIQUES

A 4-way switch cannot control lighting from four different locations as the name implies. A 4-way switch by itself cannot do much of anything. What it can do is reverse a DC motor which will be discussed further on page 52 and allow lighting to be controlled from three or more locations when used in conjunction with 3-way and other 4-way switches. Basically a 4-way switch is (2) 3-way switches tied together with internal jumper wires as depicted in figures 6-A and 6-B.

As discussed previously in "UNDERSTANDING 3-WAY SWITCHES", it takes (2) 3-way switches to control lighting from two different locations. To control this same lighting from a third location, a 4-way switch must be added. If it's desired to control this same lighting from yet an additional location, simply add one more 4-way switch. For every added location, another 4-way switch must be inserted into the circuit. For example, a light controlled from four different locations would require (2) 4-way switches and (2) 3-way switches. There will never be more than (2) 3-way switches in any conventional switch loop but the number of 4-way switches is unlimited. As shown in figure 6-C, the (2) 3-way switches are always at the opposite ends of the switch loop while the 4-way switches are all located between the 3-ways. The 4-way switch is basically a double pole, double throw switch with internal jumpers connected inside. Its job is to reverse the state of the travelers going out. No matter where the 4-way switch is located in the circuit, changing its position from up to down or vice versa will reverse the state of the travelers leaving the switch. By reversing the state of the travelers, the state of the light fixture is also reversed. What was on is now off and what was off is now on.

A 4-way switch generally has it's terminals unmarked and sometimes gets mis-recognized as a 2-pole switch because their looks are very similar. They both have four terminals but work in very differing ways. In order for a 4-way switch to function properly, it is extremely important to wire the 4-way switch into the circuit correctly. Generally, the two travelers coming in will be terminated on the top of the switch while the two travelers going out will terminate on the bottom. The same colors go on each side of the switch. Be very careful not to mix the travelers by having the black from the incoming on the top and the red from the outgoing on the top while the black from the outgoing and the red from the incoming are connected on the bottom. If this happens the switch will not operate properly.

As depicted in diagram 6-D, interrupting the 3-wire cable enclosing the black and red travelers along with the additional white wire at each 4-way switch location is relatively simple. The white wire coming in is directly spliced to the other white wire going out while the black and red travelers are connected to the four terminals on the 4-way switch. It makes good practice to slightly twist the black and red travelers in each cable together. This procedure makes it easy to identify incoming and outgoing pairs of travelers when attaching them to the terminals of the 4-way switch. It also prevents wiring mishaps and errors. Also remember that if the white wire is being used as a hot conductor rather than as a neutral conductor, it must be re-colored at all areas wherever it leaves its outer jacket cable assembly. Again, use black or colored tape other than green or white to easily accomplish this.

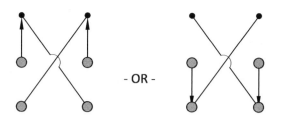 - OR -

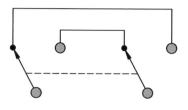

JUST LIKE (2) 3-WAY SWITCHES CONNECTED
TOGETHER WITH INTERNAL JUMPERS

6-A

• - INTERNAL TERMINALS

⬤ - EXPOSED TERMINALS

6-B

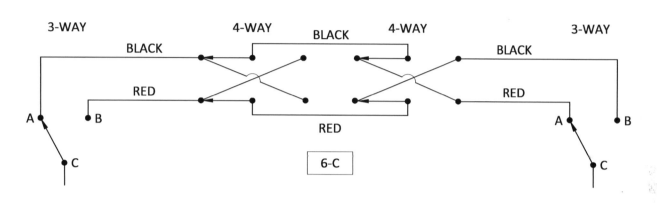

6-C

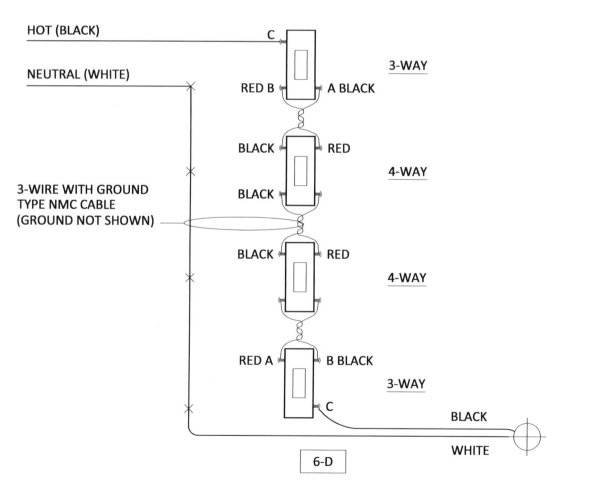

6-D

19

UNDERSTANDING MULTI-POLE SWITCHES

Multi-pole switches are considered either 2-pole or 3-pole wiring devices. They may be used to control power to two pole, three pole or multi pole loads. Multi-pole switches are generally used to control 208/240 volt, 2-pole loads or 3-phase loads. They are like having two or three individual switches under one toggle. When used in unconventional manners along with 3-way and 4-way switches, they can make tricky switching easy. Figure 7-A shows the electrical symbol and diagram for a 2-pole switch while figure 7-B details a 3-pole switch. A 2-pole switch can easily be found in most hardware and home improvement stores while a 3-pole switch may be more difficult to locate. Sometimes they can only be found at an electrical supply house. In many cases the electrical supply house will not have one in stock and it will have to be special ordered as they are not frequently used.

A conventional use of a 2-pole switch would be to control the power going to a 240 volt plug in air-conditioner or other 240 volt appliance as shown in figure 7-C. It may be used as a local disconnecting means where permitted by the code. Another common use would be to control two circuits or a multi wire branch circuit supplying a large number of high wattage light fixtures in the same room as shown in figure 7-D. If all lights are desired to be illuminated at the same time, a 2-pole switch might be considered more preferable than utilizing (2) single-pole switches. Remember however, the National Electrical Code (NEC) has a specific rule when dealing with maximum voltages allowed on adjacent switches. This rule may also apply to multi-pole switches served by different circuits when used in the same fashion. Although this rule would not normally apply on residential circuits consisting of 240 volts or less, be sure to check the code. When in doubt, check with the local building department or other jurisdiction having authority.

A 3-pole switch although not commonly found can be very useful when utilized to control 3-phase loads or multi-pole switching of individual loads. While 3-phase power is not generally available in residential occupancies, it is quite prevalent in commercial buildings and industrial settings. In some instances, a 3- phase copy machine may be employed over a single-phase unit as the wiring and circuit size needed can be smaller. In cases such as this, a 3-pole switch can serve as a local disconnecting means for the copier when the circuit breaker panel is located in another room or area as shown in figure 7-E on the following page.

Later on in this book, the roles of multi pole switches used in tricky wiring situations will be discussed and examined. It will also be shown how they can become very useful when used in conjunction with 3-way and 4-way switches. In addition, a multi-pole switch that can seemingly "trick" the travelers on a 3-way switch loop employing specialized lighting control will be discussed.

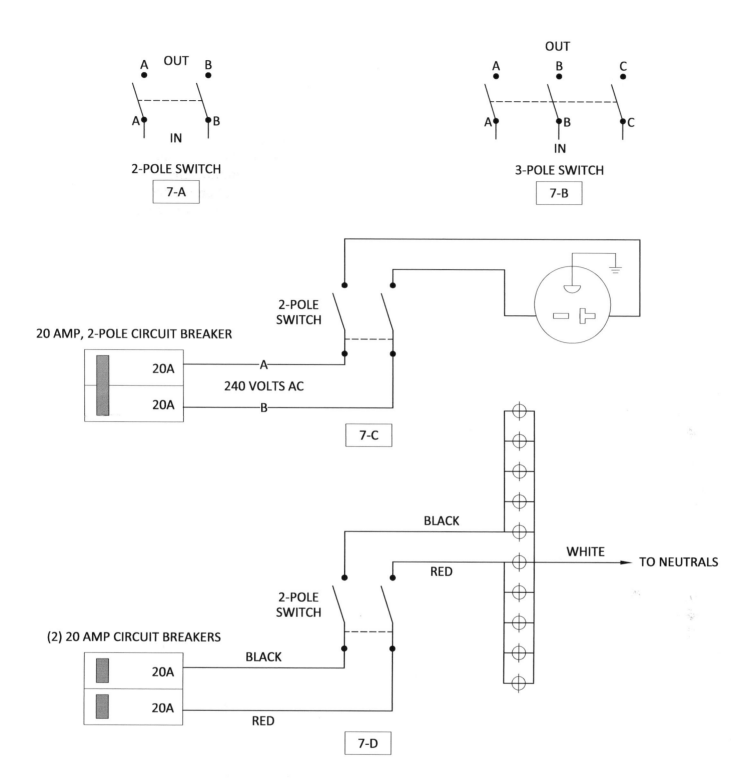

A OUT B

A B

IN

2-POLE SWITCH

7-A

OUT

A B C

A B C

IN

3-POLE SWITCH

7-B

20 AMP, 2-POLE CIRCUIT BREAKER

2-POLE SWITCH

20A

20A

A

240 VOLTS AC

B

7-C

(2) 20 AMP CIRCUIT BREAKERS

2-POLE SWITCH

20A

20A

BLACK

RED

BLACK

RED

WHITE

TO NEUTRALS

7-D

CIRCUIT BREAKER IN OTHER ROOM

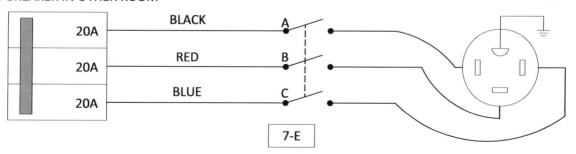

20A

20A

20A

BLACK

RED

BLUE

A

B

C

7-E

21

Chapter 2
TRICKY SWITCHING

OVERRIDE SELECTED EXTERIOR LIGHTING FROM ONE LOCATION

OVERRIDE SELECTED EXTERIOR LIGHTING FROM TWO LOCATIONS

OVERRIDE SELECTED EXTERIOR LIGHTING FROM THREE OR MORE LOCATIONS

OVERRIDE COMMON EXTERIOR LIGHTING

CONTROL SELECTED NIGHT LIGHTING FROM ONE LOCATION

CONTROL SELECTED LIGHTING FROM TWO LOCATIONS
WITH NIGHT LIGHTING FROM ONE LOCATION

CONTROL SELECTED LIGHTING FROM THREE OR MORE LOCATIONS
WITH NIGHT LIGHTING FROM ONE LOCATION

CONTROL SELECTED LIGHTING FROM THREE OR MORE LOCATIONS WHILE
CONTROLLING THREE NIGHT LIGHTS FROM THEIR OWN INDIVIDUAL LOCATIONS

3-WAY SWITCH USED AS A SIMPLE TRANSFER SWITCH

KNOB AND TUBE STYLE 3-WAY SWITCHING

EXISTING OUTBUILDING LIGHTING CONTROLS AND
DERIVING 120 VOLT POWER BY USING AN AC "OR" GATE

4-WAY SWITCH USED AS DC MOTOR REVERSING SWITCH

CUSTOM KITCHEN LIGHTING CONTROL USING 2-POLE SWITCHES

OVERRIDE SELECTED EXTERIOR LIGHTING FROM ONE LOCATION

Overriding selected exterior lighting becomes a very simple task when employing 3-way switches in an unconventional manner. Imagine the possibilities or reasons why one might want to turn on selected or all exterior lighting from the master bedroom as shown in diagram 8-A. Consider lying in bed and hearing strange noises outside the window as a particularly scary example.

Sure, this can be done utilizing proprietary wiring devices such as X-10 ®[3], and these devices may have to be used on an existing wiring job. Realize however that this can become quite expensive, especially if the designer style wiring devices and controllers are used. If installing new wiring in a custom home or room addition, simply install hard wiring with 3-way switches to accomplish this rather seemingly complicated switching task. In the following paragraphs, read on to be amazed at how simple a task this becomes and wonder why this type of wiring is not generally taught in the trade. After learning these techniques, decide whether to spread the knowledge to other electricians, friends and colleagues or keep it a tight secret.

In figure 8-B, all exterior light fixtures are connected to the common terminal on all 3-way switches. In this scenario, 3-way switches are employed rather than single-pole switches. In addition, a 3-conductor type NMB cable is run between all of the switch boxes controlling the exterior lighting. This 3-wire cable also runs between the first overridden switch location and the area where the override control is desired. Usually, this override control switch would be located in the master bedroom or other convenient location, possibly adjacent to the bedroom nightstand. By turning the override switch on, power will be provided to all exterior light fixtures that are turned off, illuminating all. Individual control of each exterior light fixture is still provided by using a 3-way switch at each location. If an individual exterior light fixture is turned on, the override switch will have no effect on that particular fixture. When the override switch is toggled to the on position it will remain on and all other exterior light fixtures that are currently off will illuminate. The correct wiring needed for the 3-conductor cable, the 3-way switches and the override switch is illustrated in figure 8-B. The power enters the override switch box containing a single-pole switch. The black hot conductor is connected to one terminal on the single-pole switch and also spliced to the black wire in the 3-conductor cable that runs between all the exterior light switch boxes. The red conductor of the 3-conductor cable is connected to the other terminal on the single-pole override switch while the remaining white conductors are spliced together. At each exterior light switch location, the black wires are spliced together and pigtailed. They are then connected to terminal B on the 3-way switch. The red wires are also spliced together, pigtailed and connected to terminal A on the 3-way switch. The black switch leg going to the exterior light fixture is then connected to the common terminal C on the 3-way switch. All of the three remaining white wires are spliced together.

This procedure is repeated at each exterior light fixture switch box that's desired to be overridden. Now individual control is provided at every exterior light fixture switch with master control at the override switch location. Simply turning the override switch to the on position as shown in figure 8-C, redirects power to all exterior lights that are off. One flick of the switch and all exterior lights are illuminated. If the override control is not desired at any particular location, simply replace the 3-way switch with a single-pole switch and cap off the unused red wire. Alternately, the 3-way switch can remain. In that case, just remove the red wire from the switch and cap it off. It's as simple as that and the hardwiring is always there. Want to add a location? Easy! Just uncap the red wire and install a 3-way switch or reconnect the red wire to the existing 3-way switch previously installed.

The genius behind this wiring method is that it doesn't have to be sold upfront or divulged that it's even there. The job can be prepped at each location using the three wire method. Later, simply cap off all un-used red wires and install single pole switches with no one being the wiser. Now you will become an electrical genius in the customers eyes when you say... "Sure! That can be done."

---

3   X-10 is a registered trademark of X-10 Ltd

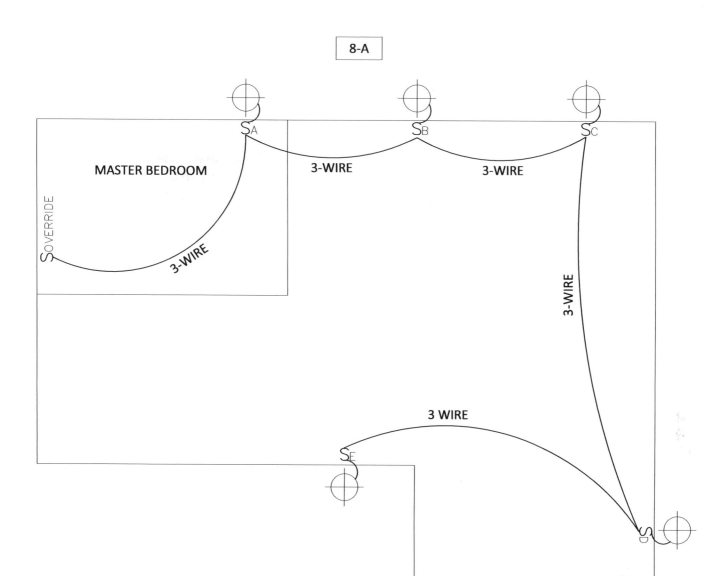

MASTER BEDROOM

S OVERRIDE

S_A   S_B   S_C

3-WIRE   3-WIRE   3-WIRE

3-WIRE

3-WIRE

3 WIRE

S_E

LEGEND

SWITCHES A, B, C, D, & E ARE 3-WAYS
OVERRIDE SWITCH IS A SINGLE POLE

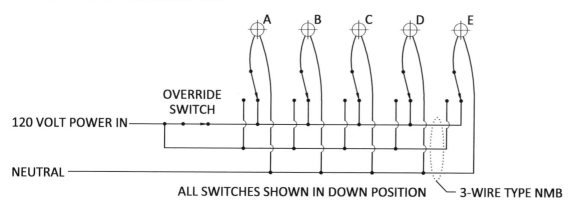

A   B   C   D   E

OVERRIDE
SWITCH

120 VOLT POWER IN

NEUTRAL

ALL SWITCHES SHOWN IN DOWN POSITION   3-WIRE TYPE NMB

OVERRIDE CONTROL:

- IF OVERRIDE DESIRED AT 2ND LOCATION: USE 3-WAY SWITCHES FOR OVERRIDE CONTROL
- IF OVERRIDE DESIRED AT 3 OR MORE LOCATIONS: INSERT 4-WAY SWITCHES IN BETWEEN 3-WAYS

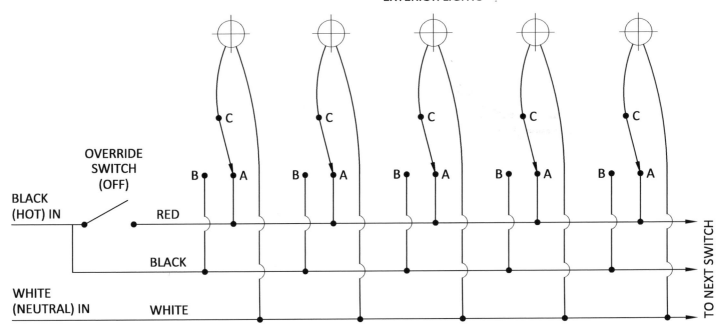

EXTERIOR LIGHTS

OVERRIDE SWITCH (OFF)

BLACK (HOT) IN

RED

BLACK

WHITE (NEUTRAL) IN

WHITE

TO NEXT SWITCH

ALL EXTERIOR LIGHT SWITCHES SHOWN IN DOWN POSITION
OVERRIDE SWITCH TURNED OFF
ALL EXTERIOR LIGHTS ARE OFF

8-B

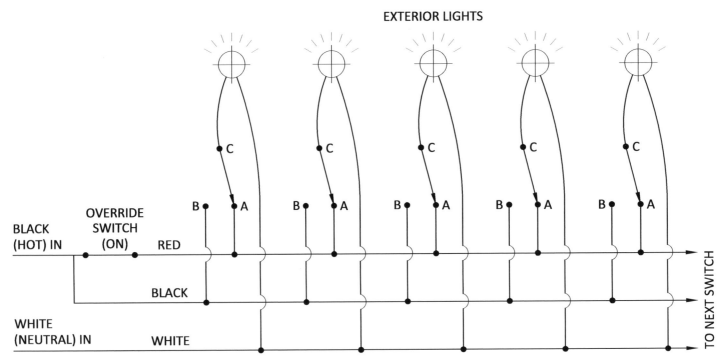

EXTERIOR LIGHTS

BLACK (HOT) IN

OVERRIDE SWITCH (ON)

RED

BLACK

WHITE (NEUTRAL) IN

WHITE

TO NEXT SWITCH

ALL EXTERIOR LIGHT SWITCHES SHOWN IN DOWN POSITION
OVERRIDE SWITCH TURNED ON
ALL EXTERIOR LIGHTS ARE ON

8-C

OVERIDE SELECTED EXTERIOR LIGHTING
FROM TWO LOCATIONS

To provide another master override switch in a second location, simply change the single-pole override switch to a 3-way switch and utilize another 3-way switch in the second desired control location. Remember, the cable running from the first 3-way override switch to the other 3-way override switch must be a 3-conductor cable. Figure 9-A shows how the power coming in as well as the override control going out must be in the same box. The other end of the 3-way override switch loop is then dead ended as learned in our "DEAD ENDING THREE WAY SWITCHES" section.

Figure 9-A depicts a set of (2) 3-way switches controlling the override function. If all exterior lights are off, changing the position on either one of the 3-way override switches will turn all exterior light fixtures on. Also, if any individual exterior light fixture locations are turned on, the override function will have no effect on that particular fixture. If it is on, it will remain on but if it is off, it will turn on. Turning any individual light on has no effect on the other exterior light fixtures. If they are off, they will remain off until they are either individually turned on or illuminated by using one of the master 3-way override switches.

This second override switch can be located at any convenient location. It can be located in the family room where everyone gathers or perhaps adjacent to the kitchen table or in a bathroom. The possibilities are endless and with a little imagination, many other locations for this override switch abound.

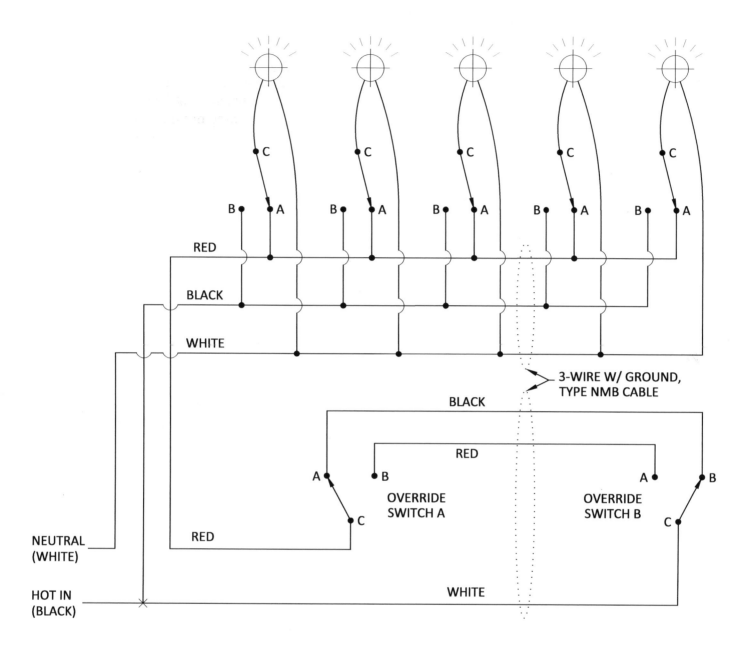

RED

BLACK

WHITE

3-WIRE W/ GROUND, TYPE NMB CABLE

BLACK

RED

A   B   A   B

OVERRIDE
SWITCH A

OVERRIDE
SWITCH B

C   C

RED

NEUTRAL
(WHITE)

HOT IN
(BLACK)

WHITE

9-A

OVERRIDE SELECTED EXTERIOR LIGHTING
FROM THREE OR MORE LOCATIONS

If override control is desired from three or more locations, inserting a 4-way or additional 4-way switches into the control loop will do the trick. Figure 10-A shows five exterior lighting fixtures being overridden at five different locations. To accomplish this feat, a switch loop containing (2) 3-way switches and (3) 4-way switches must be incorporated into the control circuit. The (2) 3-way switches are always located at each end while the 4-way switches are all located in between the 3-way switches. Take notice how the far end 3-way switch is dead ended. Also note that both the power supplying the circuit as well as the red override conductor, originate in the same box housing the closer three way switch.

Override controls may be placed in all bedrooms if desired. Perhaps the kitchen and front door areas might also be good choices. By using imagination, other areas within the dwelling may become obvious choices. With this multi-location wiring, the possibilities are almost endless.

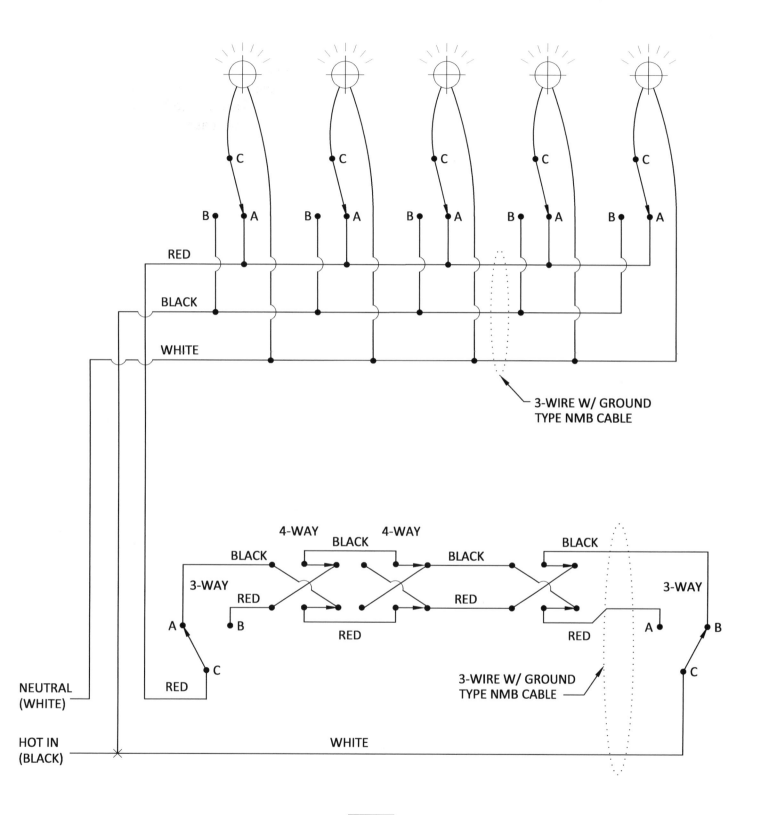

RED

BLACK

WHITE

3-WIRE W/ GROUND
TYPE NMB CABLE

4-WAY   4-WAY

BLACK   BLACK   BLACK   BLACK

BLACK

3-WAY

RED   RED

RED

3-WAY

A

B

RED

C

A

B

C

RED

3-WIRE W/ GROUND
TYPE NMB CABLE

NEUTRAL
(WHITE)

HOT IN
(BLACK)

WHITE

10-A

OVERRIDE COMMON EXTERIOR LIGHTING

Picture one of the exterior lighting fixtures being controlled from two locations such as in figure 11-A. This illustration shows two adjacent bedrooms each having exterior doors with one common light fixture between the doors. This common light fixture is controlled by 3-way switches located in each bedroom. A scenario such as this is fairly common and meets the intent of the code by providing illumination at each exterior door. The rooms need not have individual exterior fixtures so long as each door location has a 3-way switch to control the exterior light fixture.

Now for the "tricky" part. Override of the light fixture is desired but this particular scenario poses a somewhat difficult problem. There appears to be no easy solution to override this fixture along with the rest of the exterior lighting. The problem is that we are now dealing with 3-way switches at those two locations. How can these switches possibly be overridden? It's not really that hard. One way is by substituting a 2-pole switch as the override switch. The travelers can now be tricked to turn on the light along with all other exterior lights.

Figure 11-B details the wiring diagram needed to accomplish this extremely tricky task. Notice how one pole of the 2-pole switch serves as the master override control as discussed back in the first tricky switching topic. The other pole of the 2-pole switch is connected directly across the travelers in either of the 3-way switch boxes controlling the common light fixture. This second pole when connected, tricks the travelers by making them both hot at the same time. With both travelers hot, the lighting connected to the 3-way switch loop illuminates no matter what position either 3-way switch is in. If there are two locations with adjacent bedroom exterior lighting situations, substitute a 3-pole switch for the master override switch. The first pole is the normal override, the second pole tricks the first set of travelers and the third pole tricks the second set of travelers. Tricking the travelers requires an additional 2-conductor cable to be installed from the multi-pole switch location to the travelers being tricked. One pole of the 3-pole switch is required for each 3-way switch to be overridden. Figure 11-C indicates how a 3-pole switch is used to trick two sets of travelers while acting as the master override switch.

Usually, all of the exterior lighting in a single family residence, can be supplied from one 15 amp circuit. If there are too many fixtures for a 15 amp circuit, simply up the wire size to #12 and use a twenty amp circuit. Number 12 wiring may also be used on a 15 amp circuit if the total length of the conductors might possibly present a considerable voltage drop. On extremely long runs, a #10 conductor can be used if desired. The few drawbacks to using #10 wiring are the reduced flexibility, larger size and higher cost of the wire. In addition, larger capacity switch boxes may be needed to conform with the cubic inch requirements of the code pertaining to box fill.

Later in chapter 3 it will be shown how to accomplish this extremely tricky switching scenario in a second way. This can be accomplished by combining 3-way and 4-way switches in a totally unconventional manner. When multiple bedroom locations sharing a common exterior light fixture are encountered, the use of multi-pole switches is limited to two such locations. Utilizing 3-way and 4-way switches combined however, can accomplish this feat on an unlimited number of these types of locations. Rather than use multi-pole devices, consider the benefits of combining 3-way and 4-way switches.

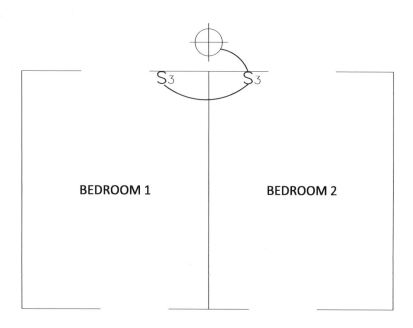

COMMON EXTERIOR LIGHT FOR 2 BEDROOMS

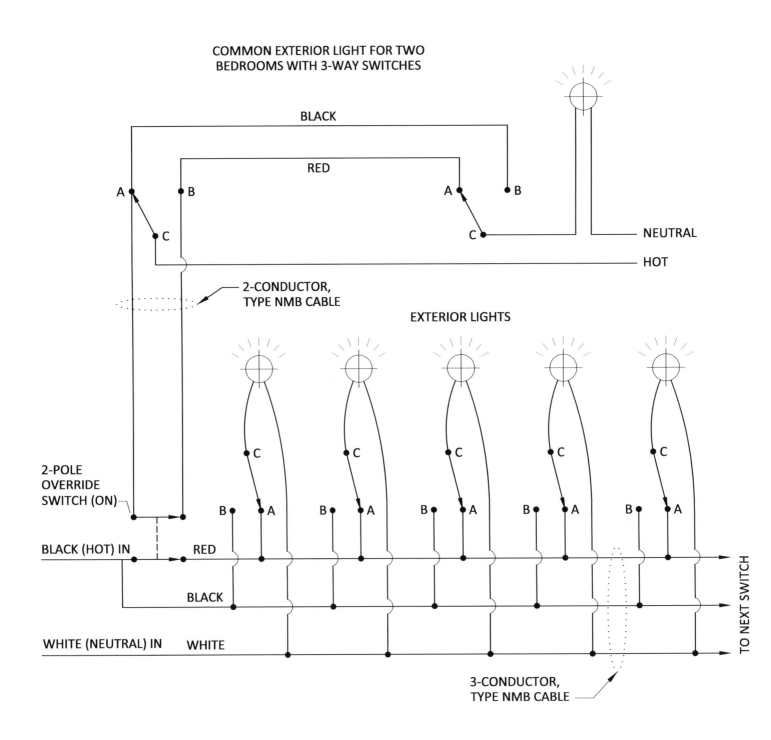

COMMON EXTERIOR LIGHT FOR TWO
BEDROOMS WITH 3-WAY SWITCHES

BLACK

RED

A   B   A   B

C   C   NEUTRAL

HOT

2-CONDUCTOR,
TYPE NMB CABLE

EXTERIOR LIGHTS

C   C   C   C   C

2-POLE
OVERRIDE
SWITCH (ON)

B   A   B   A   B   A   B   A   B   A

BLACK (HOT) IN   RED

BLACK

WHITE (NEUTRAL) IN   WHITE

TO NEXT SWITCH

3-CONDUCTOR,
TYPE NMB CABLE

11-B

34

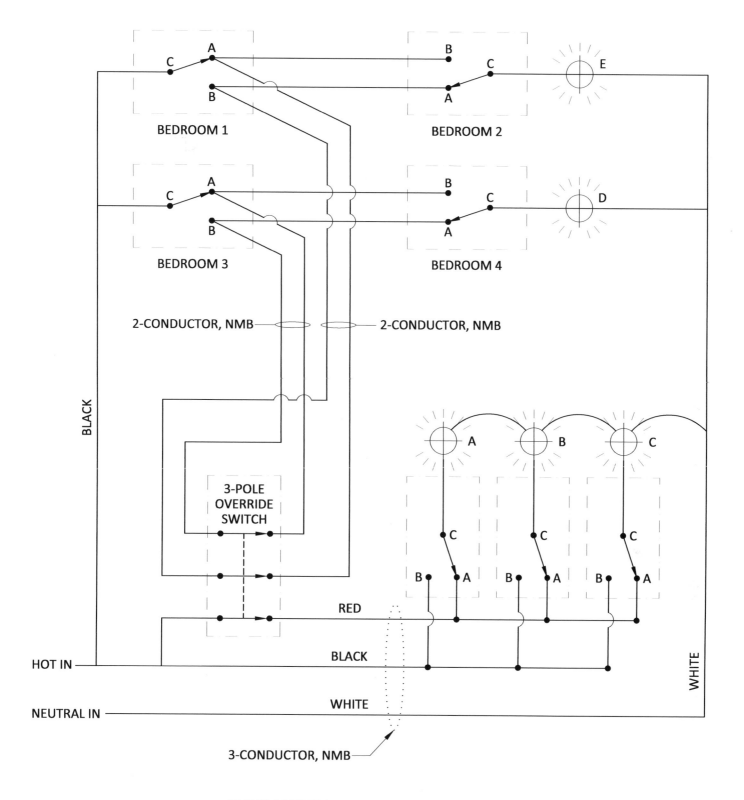

IN THIS CONDITION, ALL EXTERIOR LIGHTS ARE ON

11-C

CONTROL SELECTED NIGHT LIGHTING
FROM ONE LOCATION

In modern style residences there are usually a large number of recessed lighting fixtures in larger rooms and hallways. In many instances only selected light fixtures need be illuminated for night lighting purposes. The desired effect is accomplished by requiring only one switch to turn on all the fixtures at the same time including the night light, regardless of whether the night lighting is on or off. The night light still exhibits individual control.

Figure 12-A shows an example of a large room with ten recessed light fixtures in the ceiling. The goal is to have one fixture in the center used as a separate night light rather than illuminating the entire room at all times. This is facilitated by simply installing two switches to accomplish this task. The switches can be located side by side or remote from each other. The switch used to turn on the one night light is a 3-way switch while the other switch used to turn on all of the lights is a single pole.

Figure 12-B shows the wiring diagram needed to provide this desired illumination. Switch A is a 3-way with the common terminal connected to the individual switch leg supplying the night light. Terminal B is connected to the hot wire while terminal A is connected to the switch leg supplying the other fixtures. Switch B is a single pole connected to the remaining light fixtures in the room. Power to turn on the individual night light is available at both terminals A and B of the 3-way switch. If the switch is in the on position, power is supplied directly to the light fixture but if the switch is in the off position, power is now supplied from the switch leg serving the other fixtures. In this fashion, all light fixtures are powered on with the single pole switch while retaining individual control of the one night light fixture. If the night light is turned on, it remains on when the other switch turns on the remaining lights. If the night light is turned off, it will illuminate with the other lights when switch B is turned on.

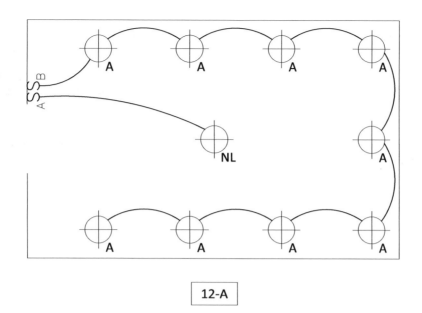

12-A

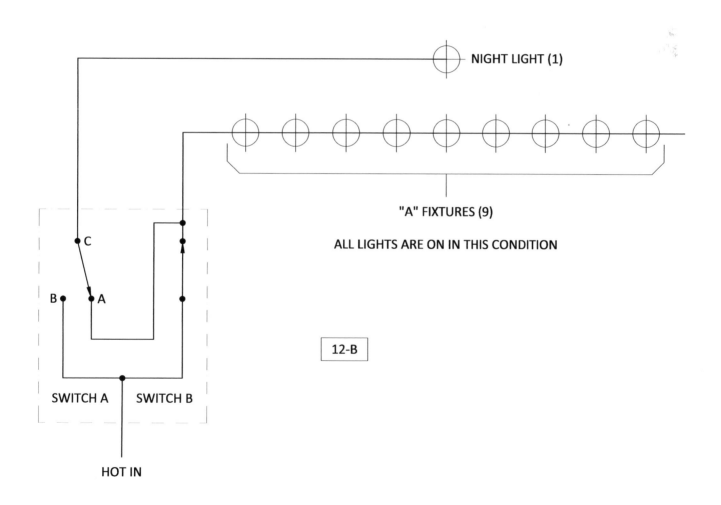

NIGHT LIGHT (1)

"A" FIXTURES (9)

ALL LIGHTS ARE ON IN THIS CONDITION

12-B

SWITCH A | SWITCH B

HOT IN

CONTROL SELECTED LIGHTING FROM TWO LOCATIONS
WITH NIGHT LIGHTING FROM ONE LOCATION

Figure 13-A shows the same large room with ten recessed light fixtures in the ceiling. In this instance there are two doorways to the room. Now, 3-way switches are located at each door to control the lighting in that room. In addition, another 3-way switch at one door is used to control the night light in the center of the room. This lighting scenario is the same as the one on the previous page excepting that the main lighting is now controllable from two separate locations.

Figure 13-B shows the electrical schematic needed to control the main lighting along with the other individual 3-way switch used to control the night lighting. The night light switch leg is connected to the common terminal C of the individual 3-way switch. Terminal B is connected to the hot wire and terminal A is connected to the switch leg serving the other nine light fixtures. This wiring technique insures individual control over the night light while still allowing the override function desired. When the night light switch is in the off position, toggling either one of the 3-way switches which are connected to the other nine lights will illuminate the night light also. When the night light is on, it will remain on when the other lights are turned on. If the night light is off, it will illuminate with the other nine lights whenever they are toggled to the on position. Either three way switch controlling the main lighting will override the switch controlling the night light.

Later in chapter 3, it will be shown how by combining 3-way and 4-way switches, night lighting can be controlled from more than one location.

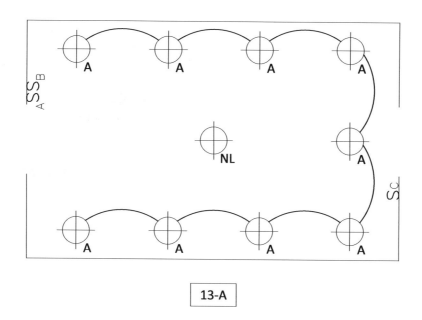

13-A

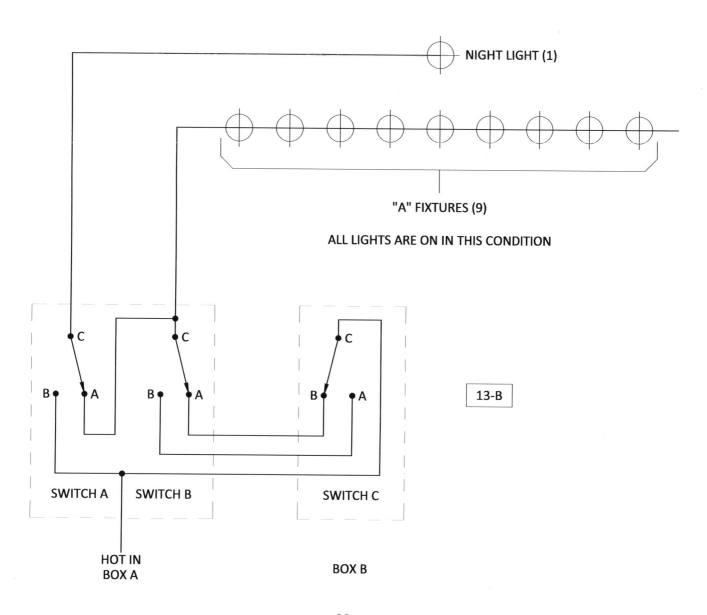

NIGHT LIGHT (1)

"A" FIXTURES (9)

ALL LIGHTS ARE ON IN THIS CONDITION

13-B

SWITCH A   SWITCH B

SWITCH C

HOT IN
BOX A

BOX B

CONTROL SELECTED LIGHTING FROM THREE OR MORE LOCATIONS WITH NIGHT LIGHTING AT ONE LOCATION

On this page, figure 14-A shows a long hallway with seven recessed light fixtures. This hallway has an entry door in the center and two other openings leading to other rooms. In the center of the hallway, one recessed fixture shall be used as a night light with a 3-way switch control at that location. In addition, all lighting in the hallway will be controlled from three separate locations. One 4-way switch is located at the entry door while a 3-way switch is located at each end of the hallway. This is similar to the previous two switch scenario as shown on figure 13-B on page 39. Two 3-way switches will still be located at either end of the hallway to control the main lighting and (1) individual 3-way switch will be located across from the entry door to control the night light. In addition, (1) 4-way switch will be inserted into the main lighting switch loop. The 4-way switch will be located between the two hallway 3-way switches and enable the lighting to be controlled from a third location.

Figure 14-B shows how the 3-conductor cable is run between the 3-way and 4-way switches. The switch leg serving all of the fixtures other than the night light is connected to the common terminal of the 3-way switch at the end of the hallway. Another two conductor cable is then extended from any convenient recessed "A" light fixture and down into the switch box containing the night light switch. The lone night light switch is an individual 3-way switch with its common terminal connected to the night light. Terminal B on the night light's 3-way switch is then connected to the same power supply feeding the other 3-way switch while terminal A is connected to the switch leg coming from any of the other six recessed "A" light fixtures.

Now, there is individual control over the night light located at the entry door but this fixture can also be overridden by toggling any of the other switches in the main switch loop. Any 3-way or 4-way switch in the main loop will energize the six other lights along with overriding the night light switch and illuminating that light fixture also. If the night light switch is already on, the override switches will have no effect. The night light will remain on. If the night light is toggled off, it will illuminate with the other six recessed lights whenever any of the 3-way or 4-way override switches are toggled. To add an override switch at more locations, insert an additional 4-way switch or switches into the control loop.

Later in chapter 3, it will be shown how by combining 3-way and 4-way switches in a totally unconventional manner, night lighting control can be switched on or off from more than one location with relative ease.

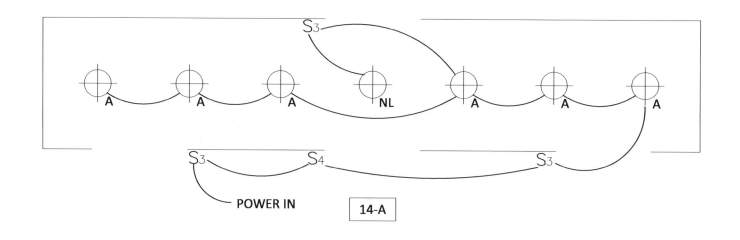

A   A   A   NL   A   A   A

S_3

S_3 S_4 S_3

POWER IN

14-A

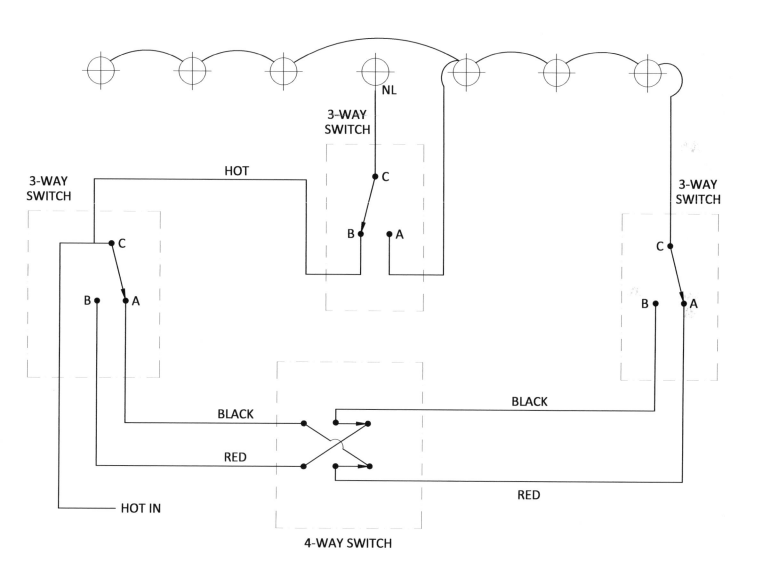

3-WAY SWITCH

NL

HOT

3-WAY SWITCH

3-WAY SWITCH

C

B   A

C

B   A

C

B   A

BLACK

BLACK

RED

RED

HOT IN

4-WAY SWITCH

14-B   ALL LIGHTS ARE ON IN THIS CONDITION

CONTROL SELECTED LIGHTING FROM THREE OR MORE LOCATIONS WHILE CONTROLLING THREE NIGHT LIGHTS FROM THEIR OWN INDIVIDUAL LOCATIONS

Here's some more cleaver lighting techniques. On page 43, the same hallway is shown with seven recessed light fixtures. One set of switches are located at the entry door and two other sets of switches are located at each end of the hallway. This time the goal is to have three night lights at three separate locations. These night lights require that their own switches be conveniently located at the doorways where the lights are located. In addition, the desired result is to be able to override the night lights from three other switch points. One location at the front door and the other two locations at each end of the hallway. For aesthetics, both the night light switch as well as the override switch will be located side by side in the same switch box at each location.

Figure 15-A shows the layout of the hallway with the recessed light fixtures and switch locations. The lighting is labeled as fixtures A, B, C and D. Fixture A is the left night light, fixture B is the center night light and fixture C is the right night light. The other four fixtures are all labeled D.

The switches will be labeled in accordance with the light fixtures that they control. Switch A will control night light fixture A, switch B will control night light fixture B and switch C will control night light fixture C. Switches D1, D2 and D3 will control the D fixtures. Switches A, B and C individually control night light fixtures A, B and C while Switches D1, D2 and D3 not only control all D light fixtures but also override the night light switches.

In this complex lighting scenario, one, two or three night light fixtures may be illuminated at any time without effecting the other light fixtures. The center night light fixture can be illuminated all by itself as can one or both of the end hallway night light fixtures. In any case, all night light fixtures will be overridden from any of the switch D locations. With any or all of the night light fixtures off, toggling any D switch will override the individual controlled night light and turn it on. If it is already on, it will remain on. If it is off, it will now illuminate with the rest of the D fixtures. Switches D1 and D3 are 3-way switches while switch D2 is a 4-way switch. Night light switches A, B and C are also 3-way switches.

Later on in chapter 3, details on controlling night lighting from additional locations will be drawn out and explained. Controlling these multiple night lights from different locations using a combination of 3-way switches and 4-way switches in a very unconventional manner will make this difficult switching arrangement very simple.

Figure 15-B shows the wiring diagram needed to control this complex switching scenario. Notice that it is not very complicated yet controls some very complicated switching. This technique can easily be varied to establish control over two different levels of lighting in the same hallway as shown in figure 16-A. In this scenario the seven lighting outlets are identified as A or B. Starting at the left light fixture we alternate until we reach the far end of the hallway (A, B, A, B, A, B, A).

Figure 16-B details how switch A will control all of the A fixtures while switch B will act as the override switch to turn on all lights, both A and B. If switch A is in the off position, switch B will turn on all the lights. If switch B is in the off position, switch A will only turn on the corresponding A fixtures. This allows two different levels of lighting with the option to override. To add override switches at additional locations, change the single-pole override switch to a 3-way switch and add another 3-way switch at the desired location. For even more additional override locations, install 4-way switches into the control loop.

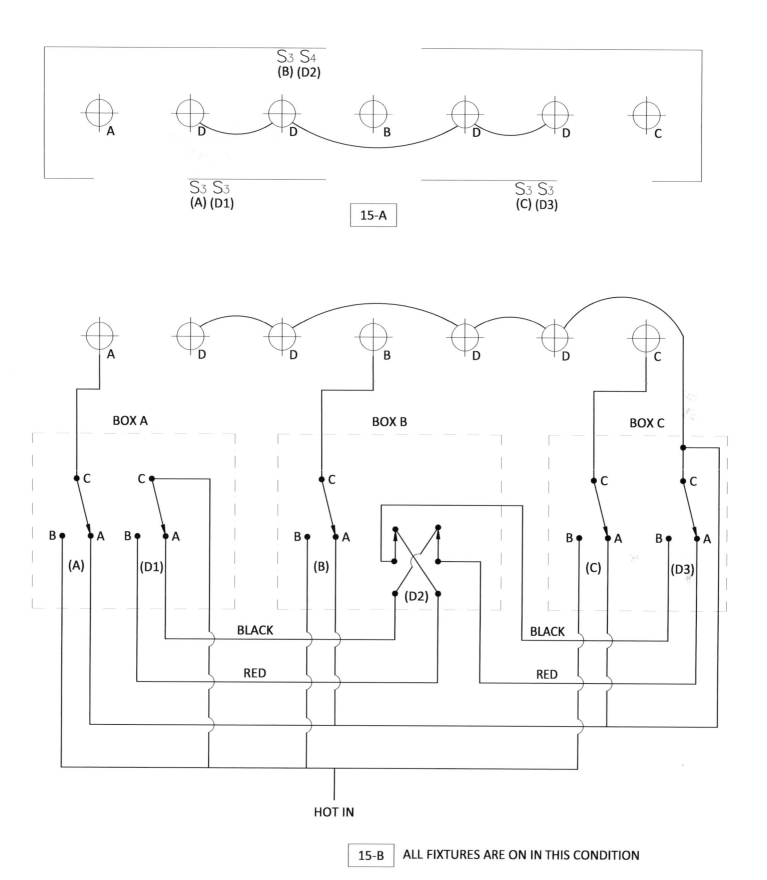

S₃ S₄
(B) (D2)

A

D — D — B — D — D

C

S₃ S₃
(A) (D1)

S₃ S₃
(C) (D3)

15-A

A   D   D   B   D   D   C

BOX A   BOX B   BOX C

C   C   C   (D2)   C   C

B   A   B   A   B   A   B   A   B   A

(A)   (D1)   (B)   (C)   (D3)

BLACK   BLACK

RED   RED

HOT IN

15-B   ALL FIXTURES ARE ON IN THIS CONDITION

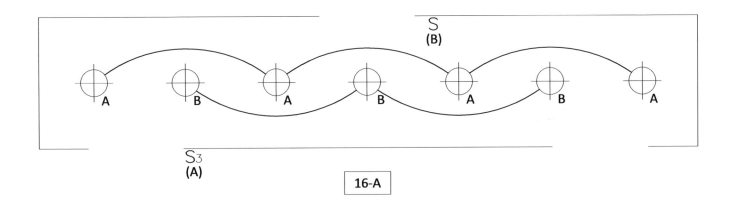

S
(B)

S₃
(A)

16-A

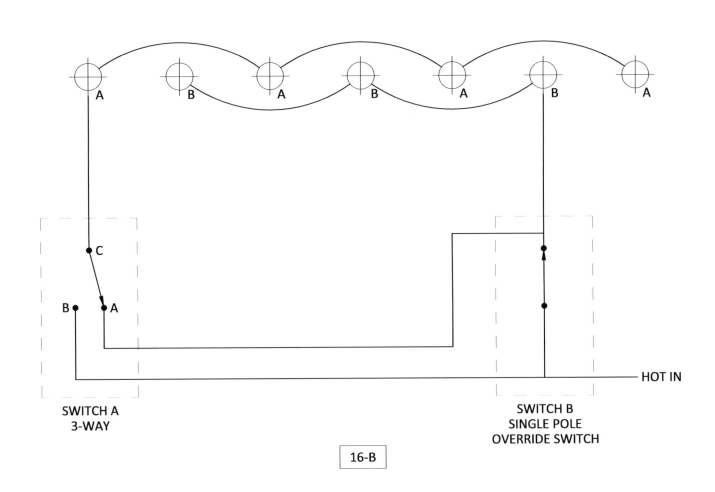

C

B •   • A

SWITCH A
3-WAY

HOT IN

SWITCH B
SINGLE POLE
OVERRIDE SWITCH

16-B

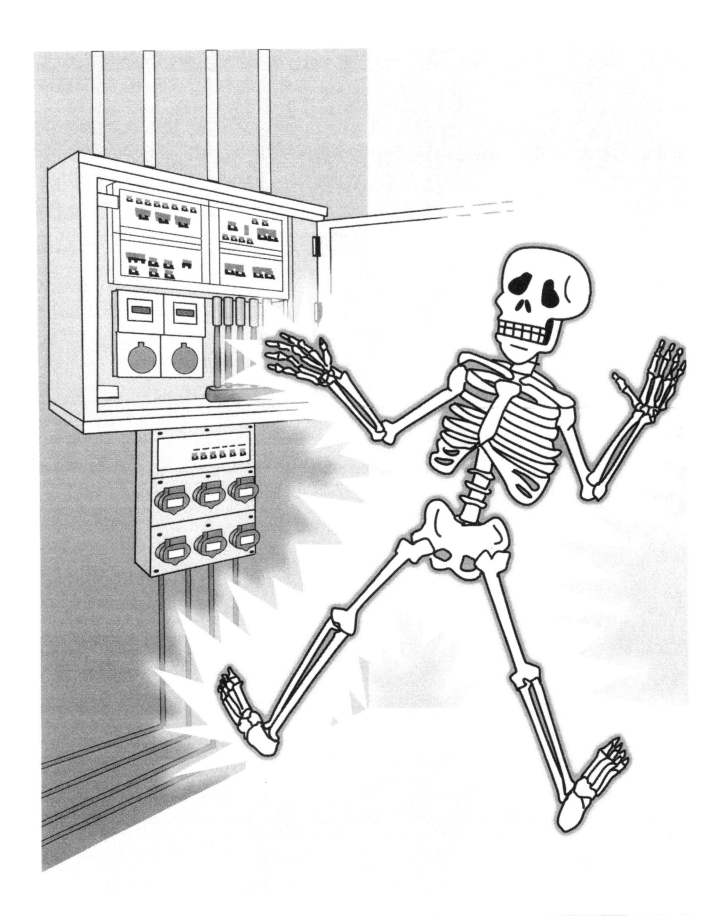

3-WAY SWITCH USED AS A SIMPLE TRANSFER SWITCH

When a 3-way switch is used as a simple transfer switch, this device will allow a light fixture, appliance or circuit to be served from two separate sources of power. One source might be from the normal utility power supplied to the house, while the other may be power supplied from a portable generator. Whenever this combination is present, extreme care must be taken not to power up the utility lines from the generator. Doing so may induce up to 12,000 volts on a normally de-energized power line; a line that utility workers may be working on. Consequently, it's mandatory to use a transfer switch when hooking up a portable generator to your house wiring. A transfer switch allows only one source of power to be utilized at a time; either the utility power or the generator supplied power. The illustration in figure 17-A shows how a standard 3-way switch functions as a transfer switch when connected as shown. The C or common terminal is connected to the load, terminal B is connected to utility power and terminal A is connected to the generator output. When connected in this manner, power to the load can be from either the utility or the generator. In the down position, the power is connected from the generator to the load via the switch contacts (C to A) while in the up position (C to B), power is supplied from the utility. Note that the power from the generator cannot be back fed into the utility line when connected in this fashion. Please note that many local jurisdictions do not approve the use of standard residential or commercial grade 3-way switches as a transfer switch. Heavier duty transfer switches with positive make/break actions are usually required. As always, check with the jurisdiction having authority before proceeding with any electrical work.

Another unconventional use of a 3-way switch is to control two different light fixtures or appliances from one source of power. On an older home there may be only one circuit supplying either a disposal or a dishwasher. The circuit is usually not adequate to run both appliances simultaneously and may frequently blow fuses or trip circuit breakers if called upon to do so. By using a 3-way switch, only one appliance can be run at a time. As shown in figure 17-B, when the switch is in the up position (C to B), the disposal will run and when the switch is in the down position (C to A), the dishwasher may be used. These two appliances when wired in this fashion, would be considered non coincidental loads. A heater and an air conditioner would be another example of non coincidental loads as they would not normally be used simultaneously.

Figure 17-C shows how to connect the dishwasher and disposal using the above method. If the old appliance is served by an armored cable such as BX, the new wiring can be fished up into the switch box and connected using a new 3-way switch. In this scenario, either a duplex receptacle or single receptacle can be used.

Fixture 17-D shows the procedure when both appliances are to be plugged in. Note how the tab on the hot side of the receptacle must be split in order to allow individual control based on the 3-way switch position.

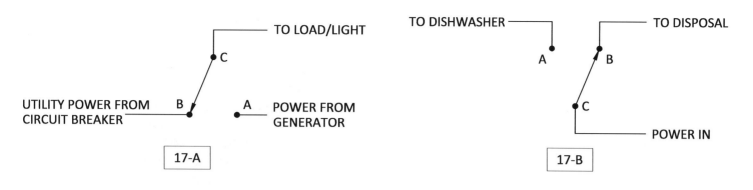

TO LOAD/LIGHT

C

UTILITY POWER FROM
CIRCUIT BREAKER

B

A POWER FROM
GENERATOR

17-A

TO DISHWASHER

A

B TO DISPOSAL

C

POWER IN

17-B

USED TO SUPPLY HARD WIRED APPLIANCES OR A
COMBINATION OF HARD WIRED AND PLUG IN APPLIANCES

EXAMPLE:

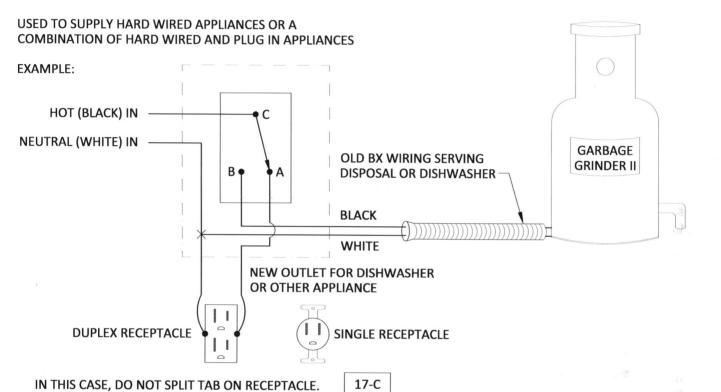

HOT (BLACK) IN

C

NEUTRAL (WHITE) IN

B A

OLD BX WIRING SERVING
DISPOSAL OR DISHWASHER

BLACK

WHITE

GARBAGE
GRINDER II

NEW OUTLET FOR DISHWASHER
OR OTHER APPLIANCE

DUPLEX RECEPTACLE

SINGLE RECEPTACLE

IN THIS CASE, DO NOT SPLIT TAB ON RECEPTACLE.
MAY USE SINGLE RECEPTACLE RATHER THAN DUPLEX.

17-C

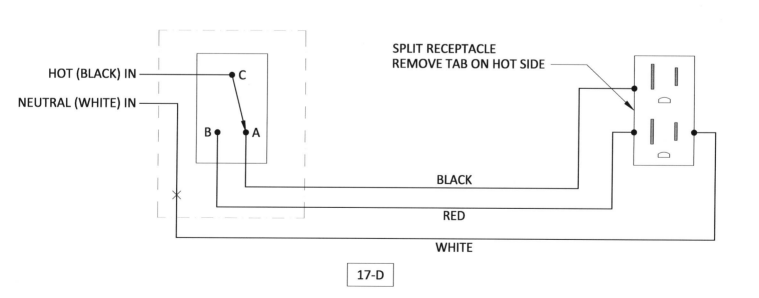

HOT (BLACK) IN

C

NEUTRAL (WHITE) IN

B A

SPLIT RECEPTACLE
REMOVE TAB ON HOT SIDE

BLACK

RED

WHITE

17-D

KNOB AND TUBE STYLE 3-WAY SWITCHING

Knob and tube style wiring was a popular wiring method employed in the 1940's and into the early 1950's. It can still be found in relatively good condition in many older homes. Older homes that have had insulation installed in the wall cavities and ceilings however, may have deteriorated wiring due to heat buildup in the wiring caused by the installed thermal insulation. The theory behind knob and tube wiring is that the conductors were basically located in free air space and could rapidly dissipate any heat generated in the conductor. Without the damaging effects of heat, the conductor insulation in many of these older uninsulated homes is still in excellent condition.

In homes that have been insulated, the trapped heat may have seriously deteriorated the conductor insulation making it very brittle in some cases. In addition, usually new electrical appliances and other loads may have been added over the years on this outdated system. This problem is further exacerbated when the circuits become overloaded. Often, fuse sizes have been dangerously increased to 30 amps to handle the added loading which further deteriorates the conductor insulation due to heat. When this style of wiring is still in use, it is best to install Type S fuse adaptors in the Edison base plug fuse socket. These adaptors limit the fuse size to the wire size and become an added safety feature which can sometimes appease insurance companies not willing to issue home owners insurance due to the older wiring and fuse problems.

3-way switches installed on knob and tube system work relatively simply but are no longer code compliant as the 3-way switches actually alternate the hot connections and neutral connections on the light fixture. Modern electrical codes require that the screw shell of the lamp socket be only connected to the neutral conductor. In this manner, a person cannot inadvertently be shocked when unscrewing or screwing a lamp into the socket while accidentally touching part of the lamps screw shell. That's because the screw shell is only connected to the neutral conductor. In Knob and Tube style 3-way systems, the screw shell may be either connected to the neutral or to the hot conductor, depending on the position of the 3-way switch.

Figure 18-A shows the wiring diagram found in these types of older homes. Notice how the hot and neutrals are connected to terminals A and B on both switches while terminal C of switch A is connected to the screw shell and terminal C of switch B is connected to the center position of the lamp socket. By alternately switching the 3-way switches, the hot and neutral conductors reverse on the lamp holder. With two hots or two neutrals connected to the lamp holder as shown in figures 18-A and 18-B, the lamp does not light. When one of the switches is changed as in figure 18-C, the lamp holder now has one hot and one neutral connected and the lamp now lights. This however is not code compliant, as the neutral may now be connected to the center of the lamp holder while the hot wire may now be connected to the screw shell. Accidentally touching the screw shell in that switch position will subject the person touching it to 120 volts. Depending on the situation, this can and will subject that person to a very serious shock, perhaps even electrocution.

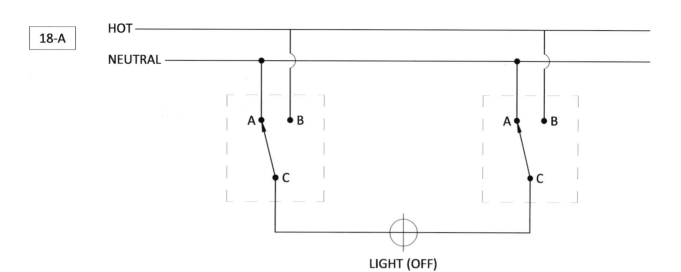

18-A

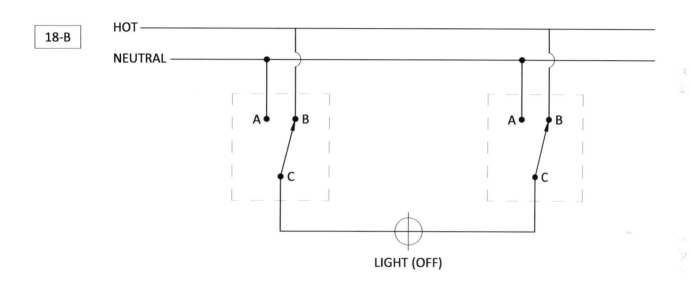

18-B

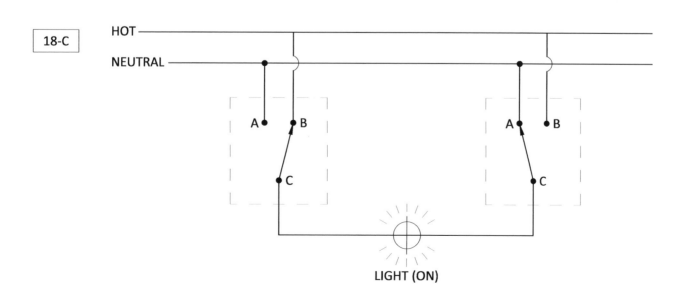

18-C

EXISTING OUTBUILDING LIGHTING CONTROLS AND
DERIVING 120 VOLT POWER BY USING AN AC "OR" GATE

Whenever a detached garage or a separate outbuilding is part of an older style wired property, there may not be any continuous 120 volt power available at the detached structure. In the olden days of wiring, the electrician simply ran three wires out to the structure to control the light fixture via (2) 3-way switches. One switch was located at the house while another switch was located at the detached garage or structure. With the light on, 120 volts was present on the traveler serving the light fixture but with the light off, there was only a neutral present on the fixture. As shown in figure 19-A, the hot conductor was only available on 3-way switch terminal B. This conductor was not connected through the switch to the fixture in this position. With this simple switching arrangement, continuous 120 volt power was not available for other electrical needs at the outbuilding location without modifying the original switching configuration.

By doing this in a legal code complying manner, the 3-way switching would have to be eliminated. In this way, lighting could be controlled at the outbuilding only, the house only or the light could be left on at all times. This arrangement however, would result in the loss of desired control, as the light could not be turned on or off from the main dwelling or at the outbuilding using 3-way switches.

Using my tricky switching scenario, there is another way around this. By constructing what I call an ac "OR" gate using simple components, 120 volt ac power will be available at all times, no matter the position of either switch. In addition, full 3-way control of the lighting fixture is also maintained. A standard dc "OR" gate is an electronic logic device that provides an output whenever voltage is present on either one of the two inputs, hence the term "OR". My definition of an ac "OR" gate is one that will provide 120 volt output power whenever there are 120 volts present on any one of the two inputs. In this case it will be the travelers.

Diagram 19-A illustrates an easily constructed ac "OR" gate using a readily available 40 amp, double-pole, double-throw, "ice cube" style relay and socket incorporating a 120 volt coil. A wire is connected from terminal B of the 3-way switch to one side of the relay coil via the socket and also to the common terminal of pole 1 on the relay. A second wire is then connected from terminal A on the switch to the common terminal of pole 2 on the relay socket. The other side of the relay coil is connected to the neutral wire. The normally open terminal of pole 1 on the relay is then jumpered to the normally closed terminal of pole 2. This jumpered wire then becomes the continuously hot wire needed for reliable 120 volt power.

By looking at the diagram, it can be seen that whenever 120 volts is present on terminal A of the 3-way switch, the relay is de-energized and the power continues through the normally closed contacts of pole 2 of the relay. If the switch position at the main dwelling is then changed, the power is removed from terminal A and is now applied to terminal B. This action will now energize the relay causing it to change state and provide power through the normally open contacts (now closed) of pole 1 of the relay. So in either state there are 120 volts present on the jumpered output of the relay. In reality, this power is not 100% continuous if the 3-way switch at the main dwelling is toggled. This is due to the fraction of a second it takes for the relay to change state. This would only happen if the switch position at the main dwelling was changed but for all practical purposes, this minuscule time period is negligible and should not affect most equipment. Changing the position of the 3-way switch located at the garage or outbuilding would have no effect on the 120 volt power at the outbuilding.

For added protection and to comply with the latest edition of the NEC, a ground fault circuit interrupter (gfci) device should be installed for any convenience outlets located in the garage. When done properly, a safe and reliable source of power is available where it did not previously exist. If desired, a properly sized ac electrolytic capacitor can be added across the output to stabilize the voltage whenever the 3-way switch at the house is toggled. Since the characteristics of a capacitor are to oppose or resist any change in voltage, the capacitor will release its stored energy back into the circuit whenever the 3-way switch at the house is toggled. In this fashion, the 120 volt power available at the jumpered relay terminal should be continuous.

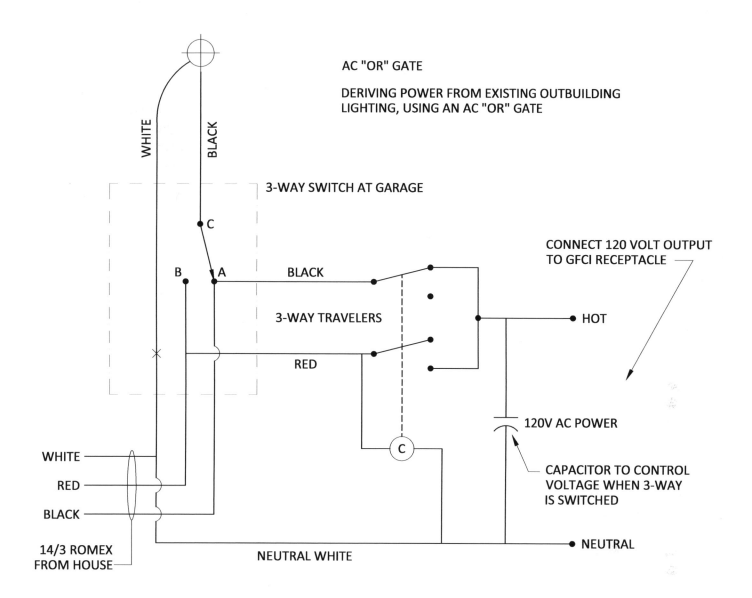

AC "OR" GATE

DERIVING POWER FROM EXISTING OUTBUILDING
LIGHTING, USING AN AC "OR" GATE

WHITE

BLACK

3-WAY SWITCH AT GARAGE

C

B   A

BLACK

3-WAY TRAVELERS

RED

CONNECT 120 VOLT OUTPUT
TO GFCI RECEPTACLE

HOT

C

120V AC POWER

CAPACITOR TO CONTROL
VOLTAGE WHEN 3-WAY
IS SWITCHED

WHITE

RED

BLACK

14/3 ROMEX
FROM HOUSE

NEUTRAL WHITE

NEUTRAL

19-A

4-WAY SWITCH
USED AS DC MOTOR REVERSING SWITCH

Most people have little reason to employ small DC motors but for the tinkerer or model builder, using DC motors can be a real plus. They have lots of speed and torque for their size and can be operated using relatively safe DC voltages such as 12 or 24 volts. Although there are special reversing switches available to control DC motors, their cost can be quite substantial. In addition, they may not be readily available through local hardware stores or even at electrical supply houses.

A simple and inexpensive residential or commercial grade 4-way switch is capable of doing the same thing. Not many people are aware of the fact that the expensive reversing switch and the inexpensive residential grade and commercial grade 4-way switches are basically identical. When looking at wiring diagram 20-A, notice how a 4-way switch is merely a double-pole double-throw switch with internal jumpers installed to reverse the state of the travelers on a lighting circuit. By using this same style switch, the positive and negative voltage needed by the DC motor can be reversed.

Figure 20-B shows the switch in one position with the positive and negative voltage entering and exiting the switch on the same side. By changing the switch position as shown in figure 20-C, it can be seen that the polarity of the voltage has now been reversed. When using this style of switch to reverse the polarity of the DC voltage coming into the motor, it causes the motor's directional rotation to change.

Figure 20-D shows how to connect a 4-way switch for proper operation. The one downside to this switching scenario is that the inexpensive 4-way switch has no center off position while the more expensive motor reversing switch does. This problem can be overcome by simply installing another single-pole or double-pole switch in series with the 4-way switching arrangement. This additional switch will now serve as the on/off switch.

Figure 20-E shows the installation of the additional single pole switch, while figure 20-F shows the two pole arrangement.

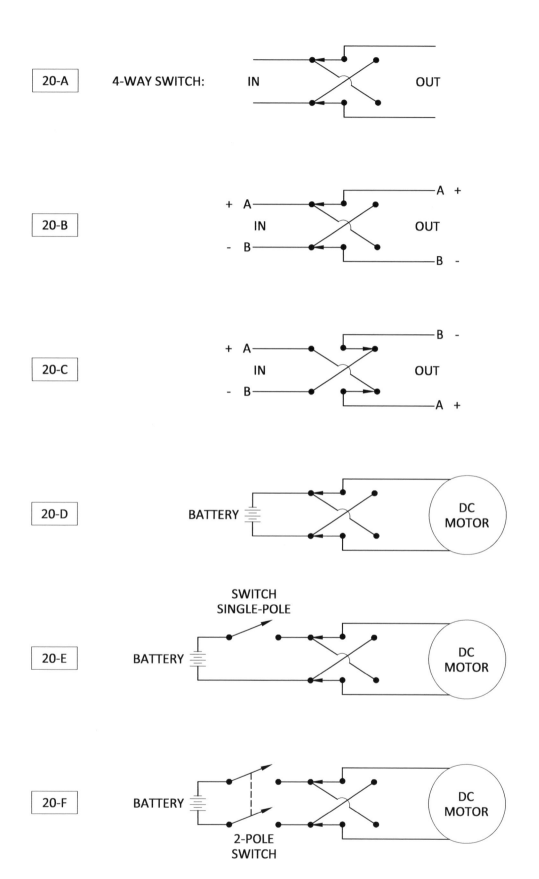

20-A   4-WAY SWITCH: IN   OUT

20-B

20-C

20-D

20-E

20-F

CUSTOM KITCHEN LIGHTING CONTROL
USING 2-POLE SWITCHES

In the pictured U shape kitchen, there are upper and lower cabinets on two adjacent walls, with the final leg of the U shape being an eating bar over the lower cabinets. The eating bar area has no upper cabinets above it. Outlining the upper cabinets are six recessed lighting fixtures. Over the eating bar are three more recessed lights for a total of nine fixtures. In the diagram on the next page, figure 21-A shows how the nine fixtures are being controlled by two separate switches. One switch controls the six outline fixtures and the other switch controls the four eating bar fixtures. Notice how the one fixture located at the end of the upper cabinets is also above the eating bar. This "common" fixture is to be illuminated whenever either switch is turned on. If switch A is toggled to the on position while switch B is off, the six recessed fixtures outlining the cabinets will illuminate. Conversely, if switch B is toggled on while switch A is in the off position, the four recessed light fixtures above the eating bar will illuminate. The "common" fixture at the end of the cabinets above the eating bar will come on whenever either switch A or B is turned on. This facilitates two levels of lighting in the kitchen with the common fixture being illuminated at each level. It allows for cabinet outline lighting or just eating bar lighting. Turning on both switches provides full illumination.

This tricky switching is accomplished by using (2) 2-pole switches to control the lighting. One pole of switch B controls the three eating bar lighting fixtures while the other pole controls the "common" fixture C. On switch A, one pole controls the five cabinet outline fixtures while the other pole controls the "common" fixture C. This is accomplished by connecting one pole on each of the two-pole switch outputs (load side) together as shown in figure 21-B. The switch leg to the common fixture is then extended to these poles. This allows the "common" fixture to become illuminated whenever either switch A or switch B is turned on. Simple to do but just as baffling to the common observer. It's almost like magic. This switching technique may be applied to any lighting configuration where one fixture is to be "common" to each switch. Add additional 2-pole switches with more sets of lights to create any kind of specialty or *"trick"* switching scenario desired. The possibilities are almost as endless as the imagination can suggest.

Another twist to this kitchen lighting arrangement can be night lighting or task lighting. Notice that one fixture is located directly over the sink. It may be desirable to have this fixture used as a night light or task light. This can be accomplished by simply installing a 3-way switch for its control. The additional switch may be conveniently located adjacent to the sink, perhaps in the same switch box housing the disposal switch. With the 3-way switch in the down position, the light above the sink would be controlled by switch A. If this fixture is to be used as a night light or task light, the 3-way switch can be toggled to the up position and only this fixture will become illuminated. Turning on switch A will override the task/night light and illuminate all A fixtures, including the one located over the sink.

In chapter 3 it will be shown how the night light or task light can be controlled from a second location by combining 3-way and 4-way switches in a very unusual manner. A manner so unusual that not even seasoned electricians know how to accomplish it.

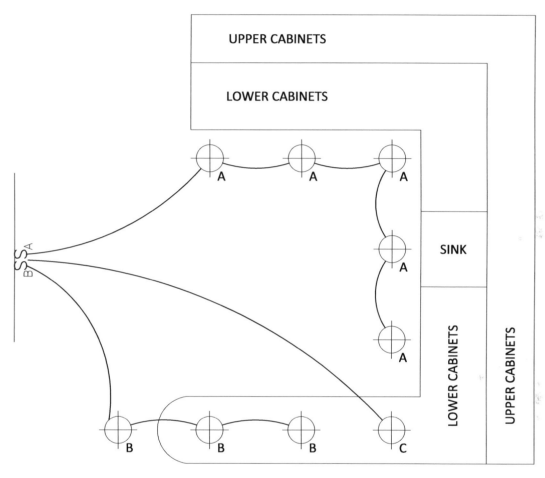

UPPER CABINETS

LOWER CABINETS

A A A

SINK

A

LOWER CABINETS

UPPER CABINETS

A

SS
A
B

B B B C

EATING BAR PENINSULA

21-A

55

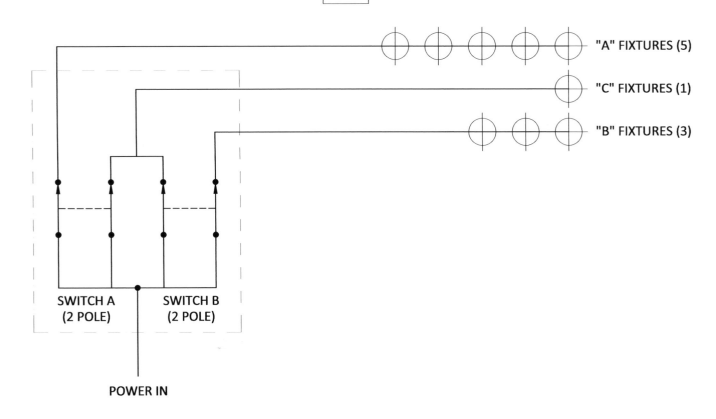

NOW WE MAKE FIXTURE "A" OVER SINK A NIGHT LIGHT:

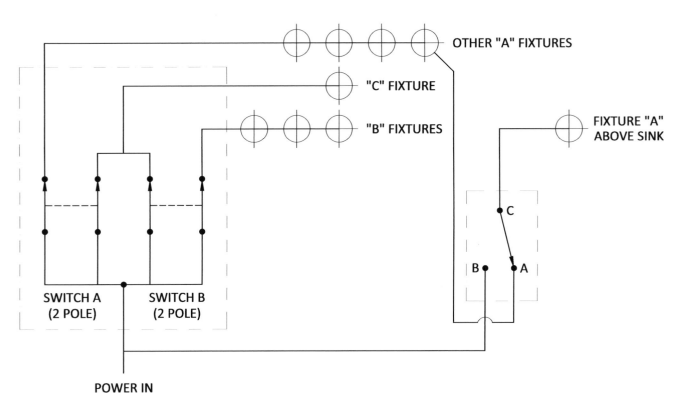

Chapter 3
EXTREMELY TRICKY SWITCHING

INTRODUCTION TO COMBINING 3-WAY AND 4-WAY SWITCHES

EXTREME OVERRIDE CONTROL USED WITH MULTIPLE COMMON EXTERIOR LIGHTING SCENARIOS

EXTREMELY TRICKY CONTROL OF NIGHT LIGHTING FROM TWO OR MORE LOCATIONS

EXTREMELY TRICKY CONTROL OF MULTIPLE NIGHT LIGHTING FROM TWO OR MORE LOCATIONS

EXTREMELY TRICKY CONTROL OF CUSTOM KITCHEN NIGHT LIGHTING
OR TASK LIGHTING FROM TWO OR MORE LOCATIONS

INTRODUCTION TO COMBINING 3-WAY AND 4-WAY SWITCHES

If someone were to approach an experienced electrician and explain that they were going to install a 3-way switch on one end while installing a 4-way switch on the other end, they would be regarded at being totally inexperienced. That's because this type of unconventional switching is not normally taught in the trade. Therefore, no experienced electrician would ever think this way. In extremely tricky switching scenarios however, this is exactly what needs to be done in order to accomplish the desired effect or control. Sure, once again this desired control can be facilitated in other ways by utilizing high tech proprietary devices. Although these specialty devices can also be used, the cost factor can become significantly higher rather quickly, so why bother. The desired effect can be reached by combining 3-way and 4-way switches together with hard wiring in a very unusual manner.

The illustrations on the following pages will detail the electrical schematics and wiring diagrams needed to utilize these devices to control extremely tricking switching scenarios. Remember that a 4-way switch when toggled, changes the state of the travelers. This unique function when combined with the unconventional use of a 3-way switch, will allow night lighting and task lighting control to be expanded to additional locations. Up until now, these locations would have been impossible to control from a second or third location.

The pages in this chapter will detail how to use this unusual combination when desiring control of night lighting or task lighting at multiple locations. Task lighting such as a fixture located above a sink in a custom kitchen can now be controlled from separate locations when employing 3-way switch and 4-way switch combinations. There can be a switch located adjacent to the sink while the other switch is grouped together with the main kitchen lighting switches. In this fashion, the task light located above the sink can be turned on or off from both locations as well as being overridden using the main kitchen lighting switches. In addition, when common bedroom exterior lighting is encountered in more than one area, utilizing combined 3-way switches and 4-way switches will achieve the same results as multi-pole switches can provide. Rather than "tricking the travelers" as learned on page 32, the 3-way switch and 4-way switch combination may be used to override these common exterior light fixtures at multiple locations in a differing fashion. Previous to learning this technique, the maximum number of common exterior bedroom lighting fixture locations desiring this control was two. Now by utilizing this combination, the number of locations controllable are unlimited. Prior to learning this, it would take a 3-pole switch to provide override control for two common bedroom locations. One pole of the 3-pole switch would be used for the override function while the other two poles would be used to *trick* the two sets of travelers. These 3-pole switches can be extremely difficult to locate so having an alternative means to accomplish the same task is an extra added benefit.

Figure 22-A illustrates the connections needed when utilizing the 3-way switch and 4-way switch combinations. In switch box A, terminal C of the 3-way switch is connected to the black wire in the 2-wire NMB cable feeding the night light or task light desiring control. Terminals A and B of the 3-way switch are connected to the black and red travelers contained in the 3-wire NMB cable which runs from switch box A to switch box B. There, the two travelers are connected to the top terminals of the 4-way switch. The two bottom terminals on the 4-way switch are then connected in the following fashion. One bottom terminal is connected to the hot conductor which allows the fixture to be controlled independently. The second bottom terminal is then connected to either the override switch or to one of the other fixtures if used for custom kitchen lighting. In this manner, the task or night light can be controlled separately from two different locations while still being overridden whenever the main lighting or override switch is toggled.

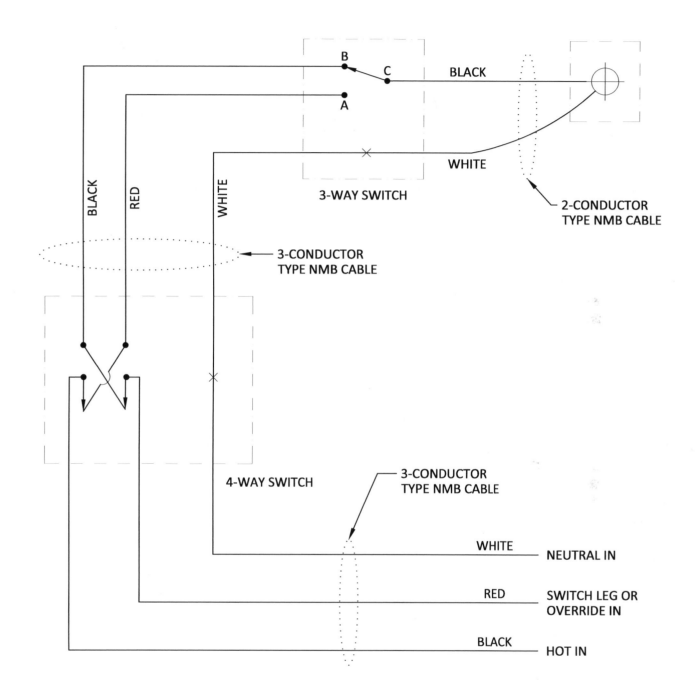

B

C

BLACK

A

BLACK

RED

WHITE

WHITE

3-WAY SWITCH

× 2-CONDUCTOR
TYPE NMB CABLE

3-CONDUCTOR
TYPE NMB CABLE

4-WAY SWITCH

3-CONDUCTOR
TYPE NMB CABLE

WHITE — NEUTRAL IN

RED — SWITCH LEG OR
OVERRIDE IN

BLACK — HOT IN

× = CONNECTION POINT

22-A

EXTREME OVERRIDE WITH MULTIPLE COMMON EXTERIOR LIGHTING

In chapter two it was shown how a common exterior lighting fixture shared by two bedrooms could be overridden by using a multi-pole switch. With one common lighting situation, a two pole switch would be used. One pole of the 2-pole switch would provide general override control to the other exterior lighting fixtures while the second pole would be used to "trick" the travelers on the common light associated with the adjacent bedroom. In another scenario comprised of two common exterior lighting fixtures shared by four bedrooms, a 3-pole switch would be employed. Again, one pole of the 3-pole switch would act as the main override control while the other two poles would be used to "trick" the two sets of travelers involved. Since the limited availability of 3-pole switches may pose a problem and also considering that the 3-pole's use is limited to two common fixtures, here's another way to do it. Combine a 3-way and 4-way switch in the following manner. In drawing 23-A, bedroom one would utilize a 3-way switch while bedroom two would use a 4-way switch. Now instead of having to "trick" the travelers, the override control is simply a function of turning on the single-pole override switch located in the master bedroom. In diagram 23-A, the power flowing through the override switch on the red conductor is connected to one bottom terminal of the 4-way switch associated with the common exterior bedroom light. The other bottom terminal is connected to the black hot wire. In this fashion, the common exterior lighting fixture is individually controlled by the 3-way switch in one bedroom and by the 4-way switch in the other bedroom. It is also overridable by turning on the override switch. If the fixture is already on, the override switch will have no effect. The fixture will remain on until either one of the bedroom switches are toggled. If the fixture is not being illuminated, toggling the override switch to the on position will redirect power over the red wire connected to the 4-way switch. From there, the power will travel through the switch and exit on the top where it is connected to the traveler connected to terminal B on the 3-way switch. That 3-way switch will transfer the power to the common light fixture that's connected to terminal C. This extremely tricky switching scenario can be duplicated at any number of common exterior bedroom lighting situations by simply running the three wire type NMB cable to the additional locations desiring control. The three wire type NMB cable is always run to the switch box containing the 4-way switch first. The 3-way switch is then connected to the 4-way switch through another three conductor cable. The switch leg serving the common fixture is connected to terminal C at the 3-way switch location. To continue this control at the second location, run the three conductor cable from the 4-way switch location to the next 4-way switch location as depicted in diagram 24-A. This can simply be repeated at additional locations where common light fixtures are employed.

Figure 24-B details the wiring schematic needed to carry out this function at four bedroom locations where each bedroom shares a common exterior light fixture. When carrying out this wiring scenario, be sure to install the cables as shown in the physical cable run diagram 24-C. The 3-conductor override cable must always be run to the 4-way switch box first. For two or more locations, the 3-conductor override cable must be run from one 4-way switch box to the next 4-way switch box location. Then install the other 3-conductor cables from the 4-way switch boxes to the 3-way switch boxes. The 2-conductor switch leg cable is run from the 3-way switch box locations to the common exterior light outlet box locations.

Figure 24-D details the wiring needed to carry out this function to additional exterior lighting locations. The 3-conductor type NMB cable carries over the hot, override and neutral conductors to all exterior lighting switch boxes. This cable is run from box to box where all individual colors are spliced and pigtailed. The conductor tails are attached to the 3-way switch serving the exterior light fixtures or the 4-way switches serving the common exterior light fixtures. Remember the 3-wire cable must run to the 4-way switch location first before running to the 3-way switch location when dealing with common exterior light fixtures. The 3-way switch location always serves the light fixture. Refer to the physical cable run diagram 24-C for clarity.

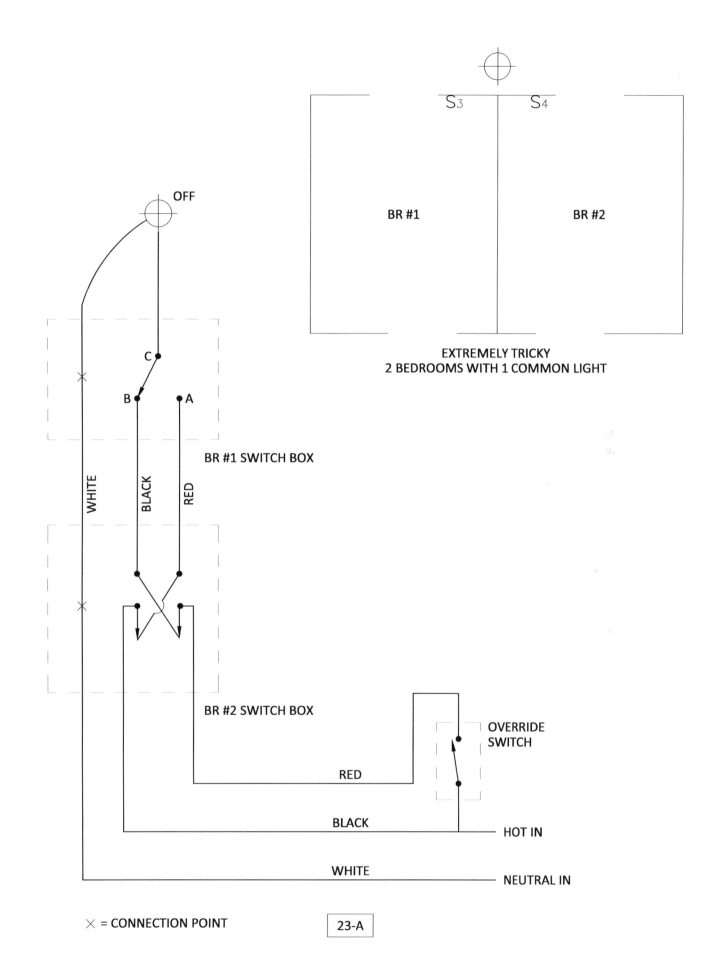

OFF

C

B · · A

BR #1 SWITCH BOX

WHITE

BLACK

RED

BR #2 SWITCH BOX

S₃   S₄

BR #1   BR #2

EXTREMELY TRICKY
2 BEDROOMS WITH 1 COMMON LIGHT

OVERRIDE
SWITCH

RED

BLACK   HOT IN

WHITE   NEUTRAL IN

✕ = CONNECTION POINT

23-A

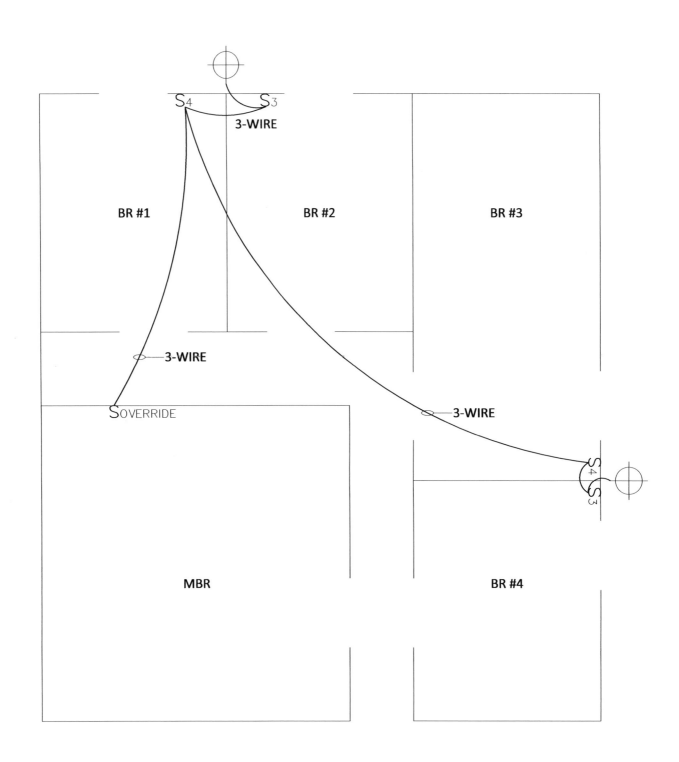

BR #1

BR #2

BR #3

S_4 S_3

3-WIRE

3-WIRE

S OVERRIDE

3-WIRE

S_4 S_3

MBR

BR #4

24-A

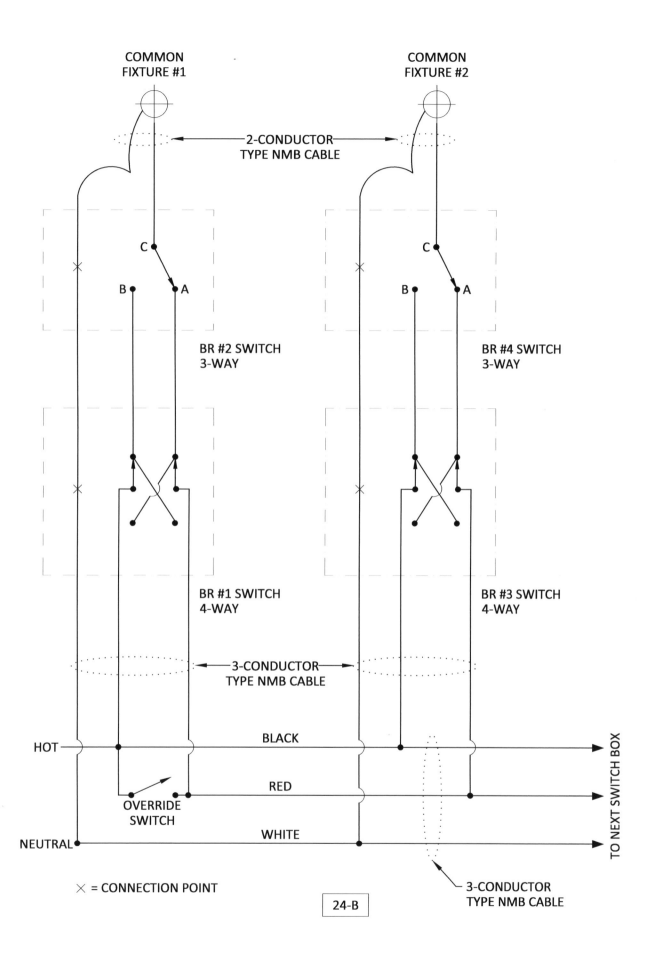

COMMON
FIXTURE #1

COMMON
FIXTURE #2

← 2-CONDUCTOR →
TYPE NMB CABLE

C

B   A

BR #2 SWITCH
3-WAY

C

B   A

BR #4 SWITCH
3-WAY

BR #1 SWITCH
4-WAY

BR #3 SWITCH
4-WAY

← 3-CONDUCTOR →
TYPE NMB CABLE

HOT ──────── BLACK ──────────→

RED ──────────→

OVERRIDE
SWITCH

WHITE ──────────→

NEUTRAL ──────────→

TO NEXT SWITCH BOX

✕ = CONNECTION POINT

24-B

3-CONDUCTOR
TYPE NMB CABLE

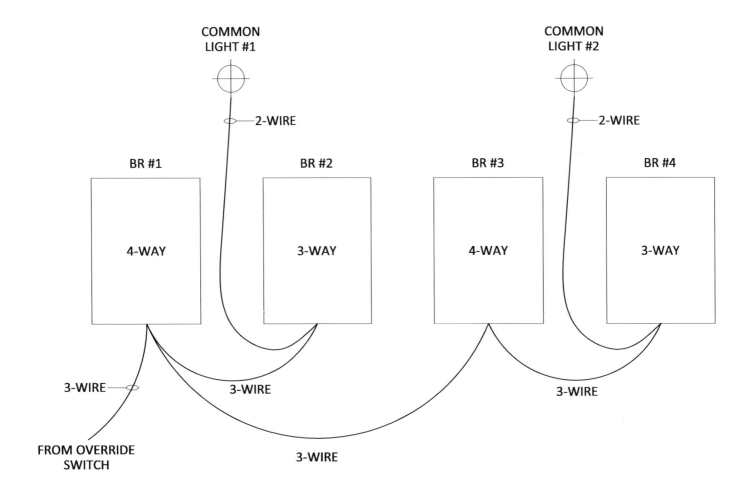

COMMON
LIGHT #1

COMMON
LIGHT #2

2-WIRE

2-WIRE

BR #1

BR #2

BR #3

BR #4

4-WAY

3-WAY

4-WAY

3-WAY

3-WIRE

3-WIRE

3-WIRE

FROM OVERRIDE
SWITCH

3-WIRE

PHYSICAL CABLE RUNS
NOTE: 3-WIRE CABLE MUST RUN TO 4-WAY SWITCH BOX FIRST, THEN CONTINUE
3-WIRE OVERRIDE CABLE TO NEXT 4-WAY SWITCH BOX. INSTALL 3-WIRE CABLES
BETWEEN 4-WAY AND 3-WAY SWITCH BOXES. ALL WHITE NEUTRAL CONDUCTORS
ARE SPLICED TOGETHER IN EACH BOX.

24-C

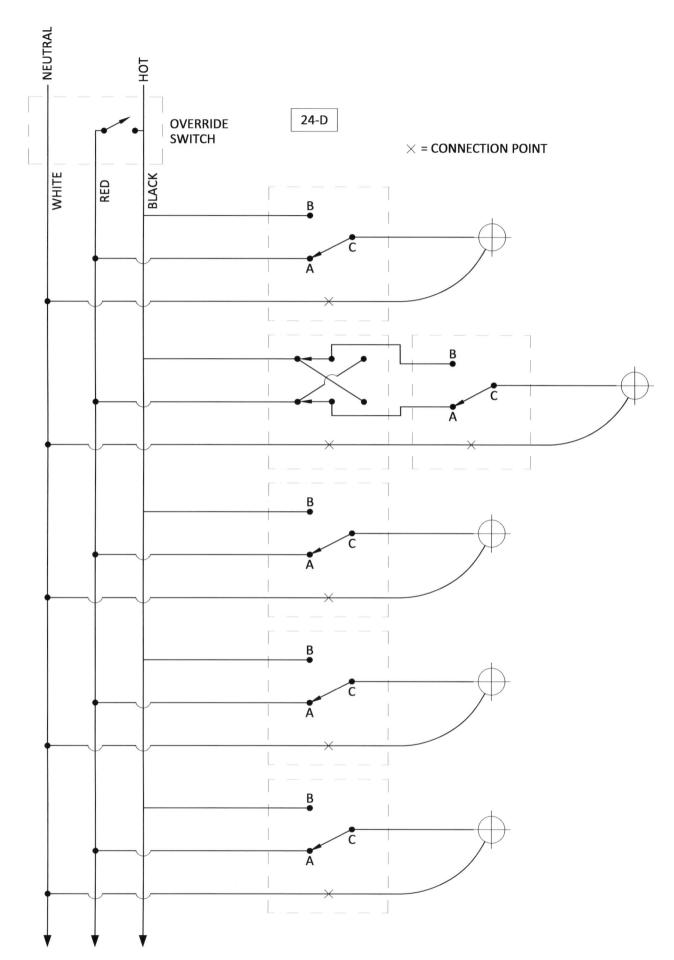

24-D

✕ = CONNECTION POINT

65

EXTREMELY TRICKY CONTROL OF NIGHT LIGHTING
FROM TWO OR MORE LOCATIONS

Controlling selected night lighting from two locations is now accomplished by combining 3-way and 4-way switches in an extremely tricky manner. In chapter two, it was shown how to control and override night lighting from one location by using a three way switch along with a single-pole override switch in an unconventional manner. It was also explained how the main room lighting was controlled at another location by employing a second 3-way switch. Now, it's desired to control the night lighting from both locations. Figure 25-A explains the technique required to control the night lighting from these two locations. Again as in chapter two, there is a large room with ten recessed fixtures and two entrances to the room. The desired effect is to illuminate the center fixture as a night light while still being able to override it whenever the main lighting switches for this room are toggled. This function is accomplished by employing (2) 3-way override switches, each located at the entry doors. In addition, the control of the night light fixture is facilitated at each switch location by employing a combination of a 3-way and a 4-way switch. This will allow individual control of the night light from both doorway locations. In this fashion, there is no need to backtrack to shut off the night light if desired. At the first doorway location, box A, a 2-gang switch box contains the 3-way override switch along with the 4-way switch needed to control the night lighting. At the second doorway location, box B, another 2-gang switch box contains the second 3-way override switch along side the 3-way night light switch. Following the diagram, it can be seen that the black hot conductor serving the room is connected to terminal C of the 3-way switch used for main lighting control. The black hot conductor is also connected to one terminal of the 4-way switch used to control the night light. With the night light fixture off, the circuit can be followed from the hot wire feeding the common terminal C on the 3-way switch used as main lighting control, through the 3-way switch to terminal B. It is then connected to one of the travelers where the power is sent over to terminal A on the second 3-way switch used to control the main lighting from the other door location. With that switch in the down position, terminal A is connected through the 3-way switch to common terminal C where it illuminates the nine "A" light fixtures as well as overriding the night light fixture. If the night light is on, the override switch will have no effect. The night light will remain on. If the night light fixture is off, the override switch will cause the night light to illuminate with the other nine "A" light fixtures. With the room totally illuminated, toggling any night light control switch will have no effect. All light fixtures will remain on.

To control this particular night light from any additional locations, simply install another 4-way switch between the 3-way switch on one end and the 4-way switch located on the other end. The 3-conductor NMB type cable will physically run from the first 4-way switch box A, to the intermediate 4-way switch box C and finally to the end 3-way switch box B as shown in diagram 25-B.

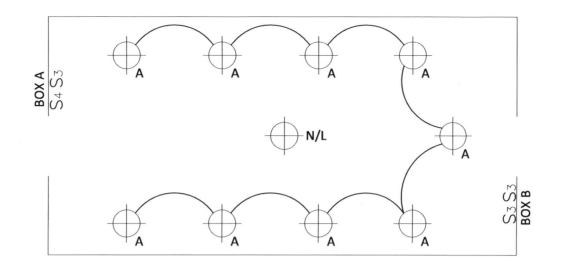

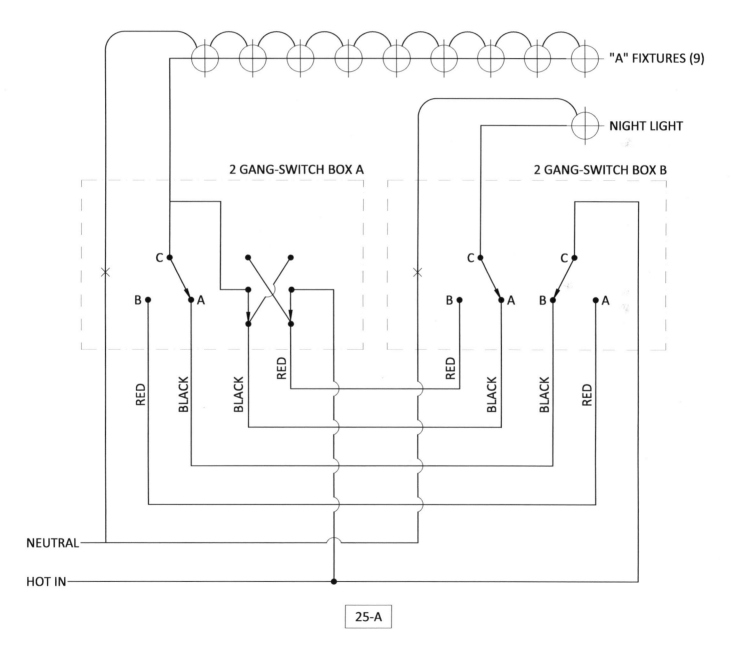

"A" FIXTURES (9)

NIGHT LIGHT

2 GANG-SWITCH BOX A

2 GANG-SWITCH BOX B

NEUTRAL

HOT IN

25-A

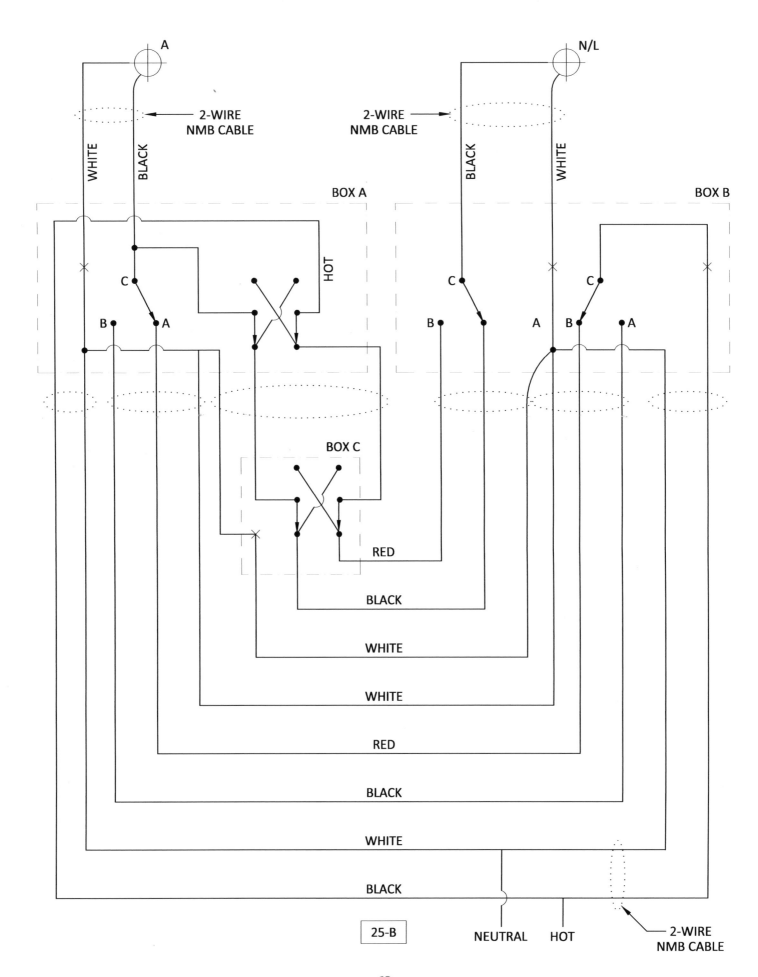

EXTREMELY TRICKY CONTROL OF MULTIPLE NIGHT LIGHTS
FROM TWO OR MORE LOCATIONS

Once again, the long hallway with seven recessed light fixtures is shown in figure 26-A. As before, there are night lights located at the entry door and at each end of the hallway. The night lights are labeled A, B and C, while all the other light fixtures are labeled D. In the past scenario described on page 42, the main lighting was controlled at three distinct locations. The first controlled location was at the front entry door. The second and third 3-way switch locations were at each end of the hallway. The night lighting was only controlled by it's own local switch or overridden by any of the main lighting switches. In this new scenario, the object is to control the night lighting from multiple points. Now the night lights can be turned on or off from another location remote from the first local switch.

Figure 26-D shows how light fixtures A and C can now be controlled from both ends of the hallway. This is accomplished by installing 3-gang switch boxes for the two hallway end boxes and a 4-way switch is added to each location. With these 4-way switches, the night lighting fixtures A and C can now be controlled from either end of the hallway. In this fashion, backtracking is not required to control the night lighting.

In the wiring diagram labeled 26-D, each added 4-way switch is connected directly to the 3-way switch serving the individual night light at the opposite end of the hallway. This is accomplished by utilizing a 14/3 type NMB cable. The black and the red travelers are connected to terminals A and B on the 3-way switches. The opposite end of the travelers are connected to the two top terminals of the 4-way switches. One bottom terminal of each 4-way switch is connected to the hot wire while the other bottom terminal is connected to the D light fixture switch leg. In this manner, whenever night light A is on or off, it can also be controlled from the opposite end of the hallway.

The same holds true for night light fixture C. Fixture C can now be controlled from either end of the hallway. Remember each each night light location is overridden by any of the fixture D switches. If any of the night lights are on, the override switch will have no effect. The night light will remain on. If the night light is off, it will illuminate with all the remaining light fixtures whenever any D light fixture switches are toggled.

If for any reason an additional night light control switch location is desired, simply install a 4-way switch between the existing 3-way switches and 4-way switches. This procedure is identical to extending any switch control to three or more locations. For each additional location, another 4-way switch is utilized and inserted into the switch loop.

Figure 26-B and 26-C detail the tricky scenario needed to control the lighting shown in figure 16-B from a third location by inserting a 4-way switch. Inserting additional 4-way switches into the switch loop will allow countless locations and endless possibilities.

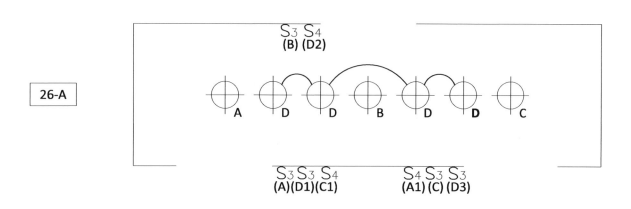

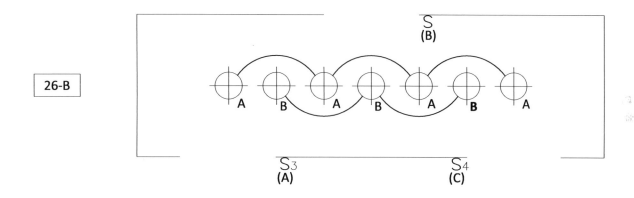

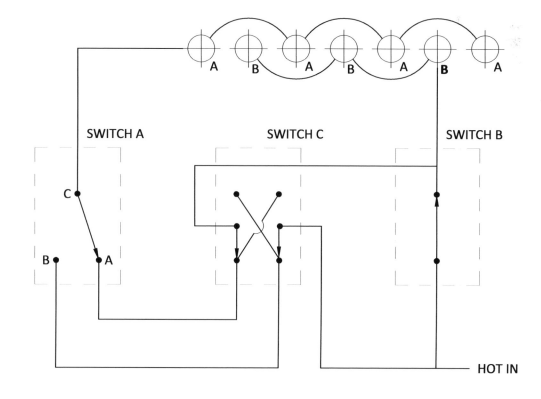

71

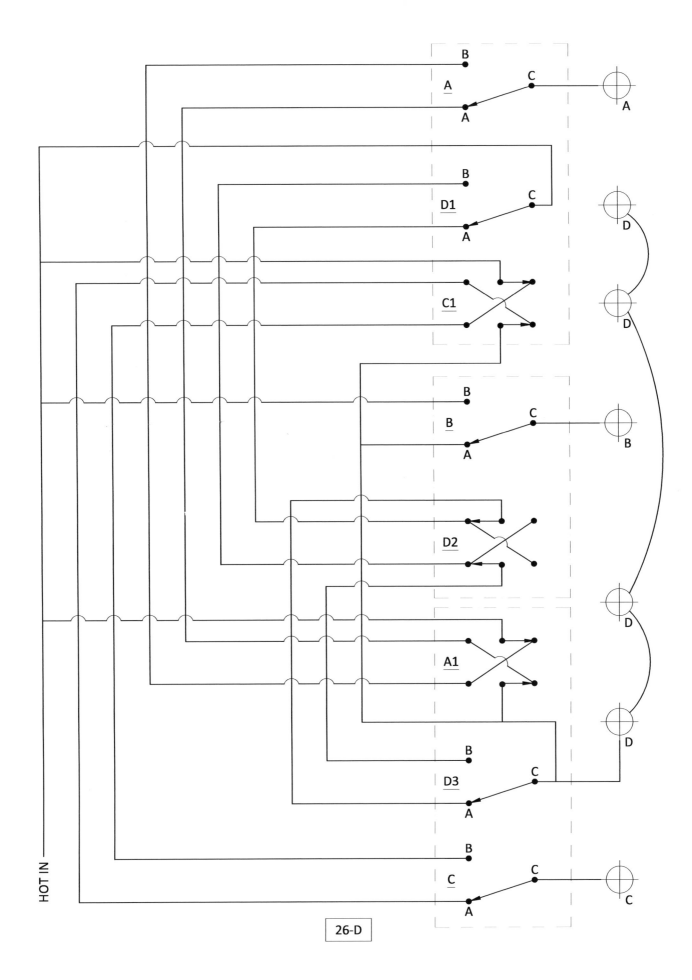

26-D

EXTREMELY TRICKY CONTROL OF CUSTOM KITCHEN NIGHT LIGHTING OR TASK LIGHTING FROM TWO OR MORE LOCATIONS

The custom kitchen cabinet lighting scenario depicted on pages 54-56 in chapter 2 can be modified to add a second location for control of the night lighting or task lighting located above the sink area. This is facilitated by combining 3-way switches and 4-way switches in a very unconventional manner.

On page 56 the night light or task light above the sink is controlled by the 3-way switch that's located adjacent to the sink. If it is desired to have an additional location, perhaps at the same location where the main lighting is controlled, a 4-way switch can be introduced into the switch loop to control this function. For ease in wiring this extremely tricky switching scenario, the 4-way switch is located in the same switch box as the (2) 2-pole switches used for main lighting control. In this fashion, the hot power wire that's needed as well as the switch leg wire for the A fixtures are already present in the same box.

Figures 27-A and 27-B detail the 3-gang switch box with the (2) 2-pole switches and (1) 4-way switch needed to control this operation. The 3-way switch is located adjacent to the sink while the 3-gang switch box is located on the wall entering the kitchen area. The hot power conductor feeds one pole of the 4-way switch while the other 4-way switch pole is connected to the output of the 2-pole switch controlling the A fixtures. The other two poles of the 4-way switch are connected via the travelers to terminals A and B on the 3-way switch located at the sink. The common C terminal of the 3-way switch is connected to the night light or the task light located above the sink. In this manner, the light fixture above the sink is controlled from two locations. One location being at the sink and the other location at the main switch box for the kitchen lighting. The light above the sink can be illuminated separately from either location. This night light fixture is also overridden whenever the 2-pole switch serving the A fixtures is toggled on. If the night light or task light is already on, the override switch will have no effect. The light will remain on. If the night light or task light is off, toggling the 2-pole switch will illuminate the night light or the task light with the rest of the A fixtures.

If the desired result is to control this night light or task light from yet another location, install a 4-way switch in the switch loop between the 4-way switch located in the main switch box and the 3-way switch box located by the sink.

This extremely tricking situation seems to do it all. There are now three levels of lighting associated with the kitchen. Level one is night lighting or task lighting. Level two illuminates all the A fixtures along with fixture C. Level three illuminates all the B fixtures along with fixture C. To accomplish this extremely tricky switching scenario without the unconventional use of 3-way switches, 4-way switches and multi-pole switches along with hard wiring techniques would be next to impossible. This special combining of 3-way switches and 4-way switches makes this seemingly impossible task a reality.

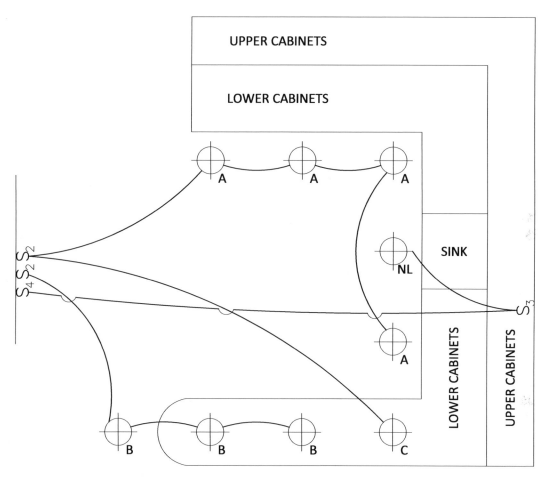

UPPER CABINETS

LOWER CABINETS

SINK

A   A   A

NL

A

LOWER CABINETS

UPPER CABINETS

B   B   B   C

EATING BAR PENINSULA

27-A

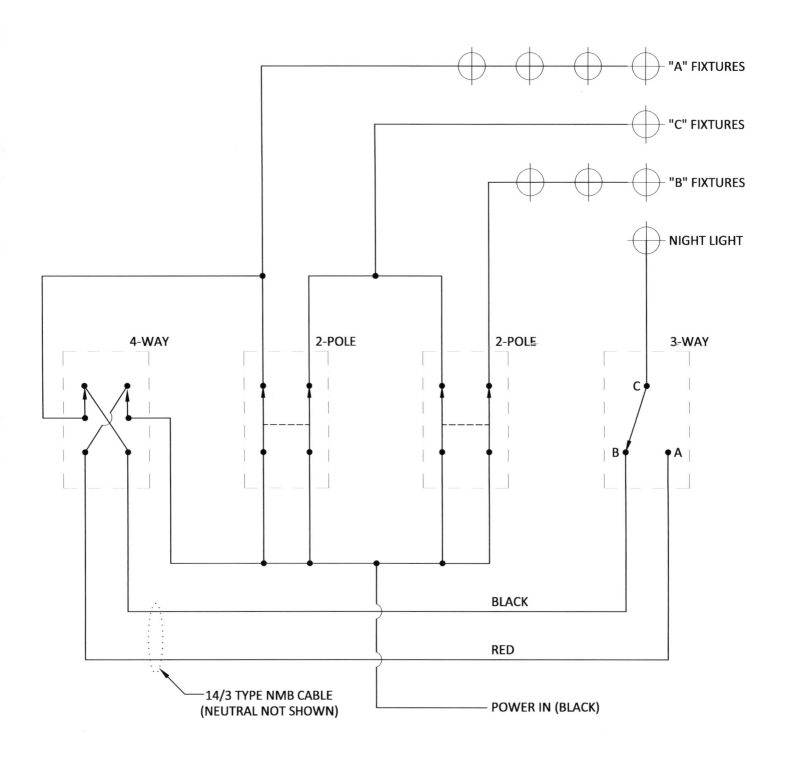

"A" FIXTURES

"C" FIXTURES

"B" FIXTURES

NIGHT LIGHT

4-WAY

2-POLE

2-POLE

3-WAY

C

B   A

BLACK

RED

14/3 TYPE NMB CABLE
(NEUTRAL NOT SHOWN)

POWER IN (BLACK)

27-B

Chapter 4
CONVENIENT FORMULAS AND OTHER USEFUL INFORMATION

THE MAGIC TRIANGLE, OHMS LAW AND POWER FORMULA

STOP!...DO NOT OPEN THAT NEUTRAL

WIRE TWISTING, IS IT REALLY NECESSARY?

EASIEST WAY TO STRIP NON-METALLIC SHEATHED CABLE

RESISTOR COLOR CODE

CALIFORNIA STYLE 3-WAY SWITCHING

WELL TANK PRESSURE CONTROLS

OTHER PROPRIETARY SYSTEMS

X-1 0 TECHNOLOGY

COMMON ELECTRICAL SYMBOLS AND DIAGRAMS

THE MAGIC TRIANGLE
OHMS LAW AND POWER FORMULA

The "MAGIC TRIANGLE" as depicted in figure 28-A, is simply an easy way to find unknown quantities in any 3-factor formula when only two of the factors are known. In drawing the triangle, the two bottom segments when multiplied together will equal the top segment. The segments are labeled as A, B and C. A times B equals C. C divided by A equals B and C divided by B equals A. The data can then be inserted into the triangle. Insert any three factor formula into the "MAGIC TRIANGLE" to find the answer. All that is required are any two factors to find the third unknown quantity.

There are many 3-factor formulas out there, with many being electrically related. Some of the more common 3-factor formulas associated with electrical work pertain to Ohms Law and power quantity. These formulas are used extensively to find voltage, current, resistance and power (wattage) quantities in electrical circuits.

When using these formulas: R= resistance in ohms
- I= current or amperage
- P= power or wattage
- E= voltage

In diagram 29-A, Ohms Law states that that I equals E over R. That is, that the current (I) in amps is equal to the voltage (E) in volts, divided by the resistance (R) in ohms. Another way to state this is that the current in the circuit is directly proportionate to the voltage and inversely proportionate to the resistance. So in a purely resistive circuit, if the voltage goes up while the resistance remains constant, the current in the circuit will increase proportionately. Conversely, if the resistance increases while the voltage remains the same, the current in the circuit will decrease inversely. Placed in the "MAGIC TRIANGLE", the Ohms Law factors would be in the triangle segments as follows: I would be in segment A, R would be in segment B and E would be in segment C. Therefore, the resistance multiplied by the current will equal the voltage. The voltage divided by the current will equal the resistance and the voltage divided by the resistance will equal the current. Place any two known quantities into the proper segments of the triangle and the answer can be calculated using simple multiplication or division.

Another useful formula is the power formula detailed in diagram 29-B. This formula is used to determine power or wattage when the voltage and current are multiplied together. Put into the "MAGIC TRIANGLE" as P equals I times E, with P being placed in the upper segment while I and E are placed into the two bottom segments. By using this formula, we can calculate that a 60 watt light bulb draws one half of an amp on regular household current. We know that the light bulb's wattage is 60 and that the household voltage is 120. This makes it easy to calculate the current using the "MAGIC TRIANGLE." Example: 60 watts divided by 120 volts equals .5 amps or one half of an amp.

So there it is. Use the "MAGIC TRIANGLE" to solve any 3-factor formula when two factors are known.

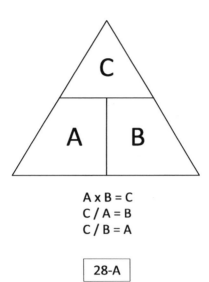

A x B = C
C / A = B
C / B = A

28-A

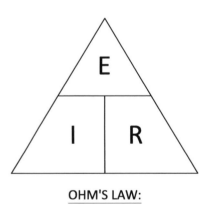

OHM'S LAW:

I = E/R

29-A

WHERE AS:
I = CURRENT IN AMPS
E = VOLTAGE IN VOLTS
R = RESISTANCE IN OHMS

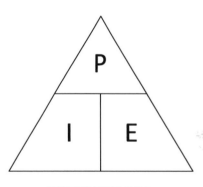

POWER FORMULA:

P = I x E

29-B

WHERE AS:
P = POWER IN WATTS
I = CURRENT IN AMPS
E = VOLTAGE IN VOLTS

STOP!...DO NOT OPEN THAT NEUTRAL

When more than one color is present in a junction box, it's a good indication that a multi-wire branch circuit is present. In that case, extra caution must be taken with the white neutral conductor. Make sure that the splices are never opened on the neutral without first having the power shut off. It's never a good idea to work on energized circuits but some electricians and handymen are either too lazy or just simply too ignorant of what can happen. Not only can a person's safety be jeopardized but an awful lot of damage can be caused to components on the circuit by opening the neutral conductor on an energized multi-wire branch circuit.

A residential multi-wire branch circuit is one that is served by two separate phase conductors sharing one neutral conductor. That means that there are (2) separate hot conductors, one from phase A and another from phase B. These two hot conductors share the same neutral conductor and would usually be contained in a single 3-conductor, type NMB, jacketed cable. The individual colored conductors in the three wire cable would be black, red and white. In addition, a forth conductor can be found in this cable. This fourth conductor is usually bare copper or colored green and is known as the grounding conductor. Although all NMB style cables referenced in this book contain the bare or green grounding wire, for the purposes of simplicity and clarity this is only being mentioned once.

The voltage between each black conductor or red hot conductor and the white neutral conductor is measured at 120 volts. The voltage between the black and red hot conductors would be 240 volts. In this type of multi-wire branch circuit the white neutral conductor only carries the difference in current between the black and the red hot conductors. If the black conductor has a load of 10 amps and the red conductor has a load of 10 amps, the white neutral conductor would carry no current and this would be considered a totally balanced circuit. On the other hand if the black conductor carried a load of five amps and the red conductor carried a load of 15 amps, the white neutral conductor would carry a load of 10 amps. This would be considered an unbalanced multi-wire branch circuit and major damage would occur if the white neutral conductor was opened during this unbalanced condition. The more the unbalance, the greater the possibility of damage. A really large unbalanced condition can be found if the black conductor was carrying a load of 16 amps while the red conductor only carried a load of 2 amps. The white neutral conductor would then carry the unbalanced load of 14 amps which is the difference between the current values in the black and red hot conductors.

To understand the magnitude of opening the white neutral conductor on an unbalanced multi-wire branch circuit, refer back to Ohms Law, the "MAGIC TRIANGLE" and the calculations on page 81. As indicated before, Ohms Law states that the voltage is equal to the current in amps multiplied by the resistance in ohms. In the large unbalanced circuit outlined above, it's know that the voltage between any hot conductor and the white neutral conductor is 120 volts. It's been further determined that the black circuit is carrying a load of 16 amps and that the red circuit is carrying a load of two amps. It can therefore be determined that the ohmic value of each circuit can be found by entering the two known factors into the "MAGIC TRIANGLE". As shown in figure 30-B, the black circuit would have an ohmic value of 7.5 ohms (120 volts divided by 16 amps) while the red circuit would have an ohmic value of 60 ohms (120 volts divided by 2 amps). Everything is fine at this point until the neutral is opened. With the neutral open, a 240 volt circuit now exists rather than two 120 volt circuits. The circuit now has a total resistance of 67.5 ohms. Again using the "MAGIC TRIANGLE", it's easy to find that the total current in the circuit is 3.555 amps (240 volts divided by 67.5 ohms) as shown in 30-C. The voltage across the loads on each circuit can be found by multiplying the current times the resistance. Circuit A now has 26.66 volts on it (3.555 amps times 7.5 ohms) while circuit B now has 213.3 volts on it (3.555 amps times 60 ohms). Guess what happens next? BOOM! Circuit B now has 213.3 volts on it instead of the normal 120 volts. Just picture that brand new flat screen television now. Of course these values are only there for an instant. Just long enough to fry whatever is on the higher ohmic value circuit. So what is learned here? First, never work on an energized circuit and secondly, make sure that all wire connections are mechanically and electrically secure, especially on the white neutral conductor of a multi-wire branch circuit.

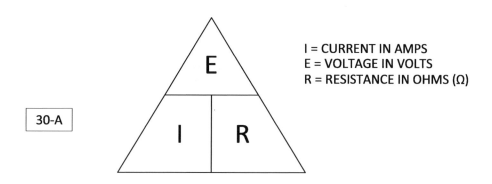

I = CURRENT IN AMPS
E = VOLTAGE IN VOLTS
R = RESISTANCE IN OHMS (Ω)

30-A

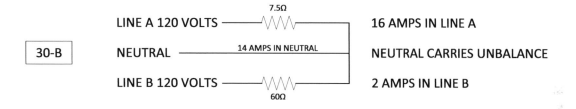

30-B

LINE A 120 VOLTS ——/\/\/\—— 7.5Ω

16 AMPS IN LINE A

NEUTRAL ———— 14 AMPS IN NEUTRAL ————

NEUTRAL CARRIES UNBALANCE

LINE B 120 VOLTS ——/\/\/\—— 60Ω

2 AMPS IN LINE B

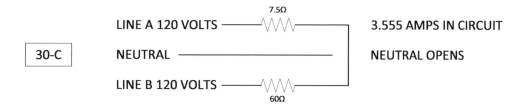

30-C

LINE A 120 VOLTS ——/\/\/\—— 7.5Ω

3.555 AMPS IN CIRCUIT

NEUTRAL ——————————

NEUTRAL OPENS

LINE B 120 VOLTS ——/\/\/\—— 60Ω

NOW WE HAVE: I = E/R   240 VOLTS / 67.5Ω = 3.555 AMPS

NOW WE HAVE: E = IxR

ON THE 7.5Ω SIDE WE HAVE: 3.555 AMPS x 7.5Ω = 26.66 VOLTS

ON THE 60Ω SIDE WE HAVE: 3.555 AMPS x 60Ω = 213.3 VOLTS

WIRE TWISTING, IS IT REALY NECESSARY?

When I was a foreman for a large residential contractor, one of the biggest complaints I received from the journeyman involved wire twisting or pre-twisting of the wires. One of my rules was that all wires had to be pre-twisted prior to installing the wire nut ®[4]. It was mandatory. Many workers thought that it was too time consuming and didn't accomplish much as the wire nuts did a perfectly good job without prior pre-twisting. They even went to the point of showing me the wire nut box and explained that the directions indicated that pre-twisting was not a requirement. I however, pointed back to the box instructions that stated that the wire nuts could be installed with or without pre-twisting.

After a while, they all eventually got use to pre-twisting and agreed that the final connection was indeed superior. The pre-twisting guarantees a good mechanical connection which further ensures a good electrical connection. Then the wire nut is basically needed only for it's insulating properties as the connection is already mechanically and electrically secure without having to rely on the wire nut spring to maintain the connection.

As far as the extra time needed to pre-twist the wires, its nothing compared to the time it would take on a callback to repair the loose connection where an individual conductor slid out of the wire nut when the bundle was pushed back into the junction or outlet box. Worse yet, remember what happens when the neutral conductor opens on a multi-wire branch circuit? Case closed! Always pre-twist. It makes for a much better installation.

To pre-twist the wires, simply leave them a little bit longer and strip off the insulation a little more. I personally like to have about one and a half to two inches of bare conductor exposed. Then line up the insulation ends and use a seven or eight inch pair of lineman's pliers to do the twisting. Grab the ends of the exposed conductors with the pliers while holding the bundle tightly together with the other hand. Then twist, release, twist, release and continue until the exposed ends look like a tightly wound spiral spring.

It's helpful if the lineman's pliers have a really loose joint to facilitate the twist, release, twist, release procedure. Whenever I purchase a new pair of lineman's pliers, I always check several pair to insure that I get the loosest jointed pair possible. You basically want the handles to drop open on their own. WD-40 ®[5]combined with a little tweaking, will usually make any pair acceptable. After the pre-twisting simply cut off the end of the twisted bundle to the required length and install the wire nut. You can also use the lineman's pliers to tighten the wire nut using the same twist, release, twist, release method employed on the wires. It may feel awkward at first but trust me, in no time it will become second nature.

---

4   Wire Nut is a registered trademark of IDEAL Industries, Inc.
5   WD-40 is a registered trademark of WD-40 Manufacturing Company

EASIEST WAY TO STRIP NON-METALLIC SHEATHED CABLE

There are many ways to strip the outer jacket off of non-metallic sheathed cable. Non-metallic sheathed cable is commonly referred to as ROMEX ®[6] or rope. There are specialty tools designed to slice the jacket down the middle or center of the cable. Most of these tools work fairly well but there's still more work to do after the slitting process. The jacket still needs to be peeled off and then cut off where the slice began.

A simpler and much cleaner method can be obtained by simply using a standard type retractable blade utility knife. Just make sure that the score is started using a brand new extremely sharp blade. Now take the knife and diagonally score the cable on the upper and lower sides using caution not to score too deeply. Try to apply just enough pressure to gently score the outer jacket without penetrating the insulation on the interior conductors. With just a little practice you'll be stripping like a pro. This procedure works quite well weather or not the cables have been inserted into the junction or outlet box. My personal preference however, is to first insert the cable into the box, staple the cable to the framing member per code, tighten any required cable clamps and then score and snap. The utility knife is real easy to get up high into the junction or outlet box so as to minimize the amount of jacketed cable left in the box. The more jacket left, the harder the wire make up procedure becomes.

The stripped jacket also makes wonderful wire markers. Just cut the stripped jacket into 1" pieces and mark the circuit names and numbers on the individual pieces with a permanent marker. These jacket pieces can then be slid back onto the cables for future identification.

This stripping procedure works exceptionally well with 2-conductor cables. Just score as directed then grasp the remaining jacket and give a sharp tug. The cable jacket instantly separates at the score mark and slides off. On 3-conductor cables the procedure is handled quite differently. The utility knife is used to gently score down the center of the cable. The jacket is then separated while pulling it upwards and away from the enclosed conductors towards the interior of junction box. Then a quick slice with the knife will finish the job. Again, diligence and dexterity must be used while handling the razor knife in order to avoid damaging the inner conductor insulation. Use only enough pressure to score lightly as this will avoid damaging the insulation on the enclosed conductors.

---

6   ROMEX is a registered trademark of Southwire Company

RESISTOR COLOR CODE

Whenever doing electrical work or tinkering with electronic hobbies, it's always a good idea to know the resistor color code. Back in high school electronics class, the instructor taught us a relatively simple way to memorize the color code. Today, this catchy little ditty may not be politically correct and might be considered offensive to some but in the interest of learning more easily, I've decided to include it anyway.

The ditty goes like this…" Bad beer rots out your guts but vodka goes well, get some now." What does that mean, you ask? Well, it's simply an easy way to remember the resistance color code. Each word of the ditty signifies one of the color bands on the resistor along with it's corresponding value. Resistors have three color bands to signify the value and one additional color band to indicate the tolerance. The chart below shows how the words correlate to the colors and values.

WORD	COLOR	1st BAND VALUE	2nd BAND VALUE	3rd BAND MULTIPLIER	TOLERANCE
Bad	Black	0	0	1	
Beer	Brown	1	1	10	
Rots	Red	2	2	100	
Out	Orange	3	3	1,000	
Your	Yellow	4	4	10,000	
Guts	Green	5	5	100,000	
But	Blue	6	6	1,000,000	
Vodka	Violet	7	7	10,000,000	
Goes	Grey	8	8	100,000,000	
Well	White	9	9	1,000,000,000	
Get	Gold				5%
Some	Silver				10%
Now	None				20%

Tolerances are plus/minus values.

Diagram 31-A explains how to decode the resistor color bands. Let's picture a resistor with four colored bands. The colored bands are red, black, orange and silver. This resistor would have a value of 20,000 ohms. The first band of red equals 2, the second band of black equals 0 and the third band of orange equals the multiplier of 1,000. Therefore, the resistor value is 20 times 1,000 or 20,000 ohms. The fourth colored band indicates the tolerance. In this example, the last silver band indicates that the resistor has a plus/minus tolerance of 10%. Since 10% of 20,000 equals 2,000, the resistor can have a value of between 18,000 ohms and 22,000 ohms. If we needed to choose a value that would be more accurate, we would consider a resistor with a fourth color band of gold. Gold represents a plus/minus tolerance of 5% so the same resistor with a gold band would have a value somewhere between 19,000 and 21,000 ohms.

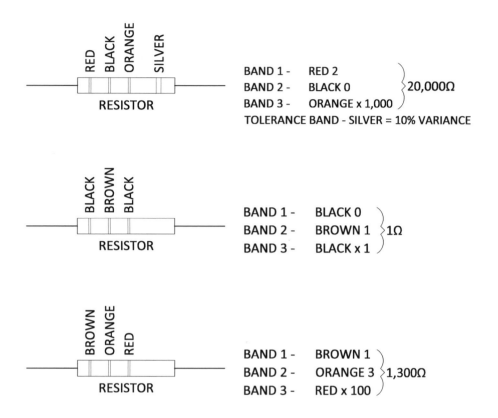

BAND 1 -   RED 2
BAND 2 -   BLACK 0   } 20,000Ω
BAND 3 -   ORANGE x 1,000

TOLERANCE BAND - SILVER = 10% VARIANCE

BAND 1 -   BLACK 0
BAND 2 -   BROWN 1   } 1Ω
BAND 3 -   BLACK x 1

BAND 1 -   BROWN 1
BAND 2 -   ORANGE 3   } 1,300Ω
BAND 3 -   RED x 100

31-A

CALIFORNIA STYLE 3-WAY SWITCHING

The drawings 32-A and 32-B show what are considered to be California style 3-way switching arrangements. Although they are fairly uncommon, they could still possibly be found in older homes today. This style 3-way uses (2) sets of 2-wire cables rather than the 3-wire cable found in most modern day structures. It was used back in the day to allow 120 volt power to be available at the second location but by today's electrical codes and standards it's considered illegal.

Many people including experienced electricians have spent countless hours trying to troubleshoot problems associated with California style 3-way switching scenarios. In many instances the home owner himself removed the existing device and opened up all the connections without marking the wires. This made troubleshooting next to impossible as all wires looked the same. Each box had at least six individual conductors so trying to put in the new switches was not an easy task. If possible, avoid ever using anything like this in modern electrical work. The diagrams shown are for illustrative purposes only and are intended solely for troubleshooting problems associated with existing California style 3-way switching techniques.

"CALIFORNIA" STYLE 3-WAY SWITCHING

NOT LEGAL, DO NOT USE
FOR INFORMATIONAL PURPOSES ONLY

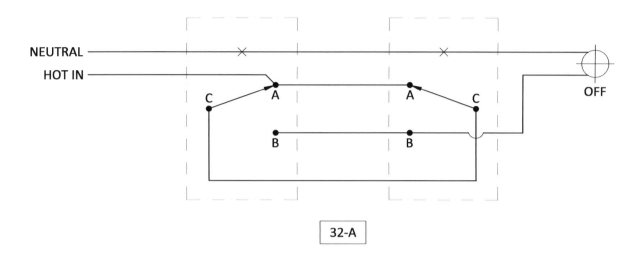

32-A

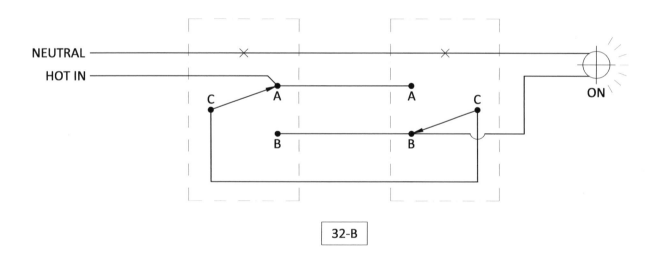

32-B

CUSTOM WELL PRESSURE TANK WIRING FOR RURAL AMERICA

Many custom homes in rural parts of the country are served by individual water wells rather than being served from a municipal water supply. One of the biggest complaints from homeowners with wells involve low water pressure. Sometimes the pressure seems to be fine but at other times it is unsatisfactory. The primary reason for the pressure fluctuations is in the way that most systems are designed. The water pressure pump and bladder tank are usually controlled by a pressure switch. This pressure switch is set to provide a range of somewhat not so satisfactory pressure constants. The switch has a high and a low pressure setting that is adjustable. Usually the high value is set at around 80 to 90 psi (pounds per square inch). The low setting may be around 40 to 50 psi. When the pressure is in the 80 to 90 psi range, the shower and other faucets seem to flow satisfactorily. Whenever the pressure drops to the 50 to 60 psi range, people using the water usually complain. The shower that once felt invigorating now feels more like it's a watering can being drizzled out. Generally the pressure will gradually return to normal during the showering period as the pressure dips to the low value and kicks the pressure pump back on. As the pressure pump continues to run, eventually it will catch up and provide higher pressure for the duration of the shower. As long as the pressure pump and bladder tank are in good condition they can usually provide decent pressure when called upon to do so.

The main problem is that the pressure pump will not kick back in until the water pressure has dropped to the preset value on the low side of the pressure switch. Starting the shower when the pressure is high gives good water pressure at first but then gradually reduces as the shower progresses. The longer the shower, the longer the cycles of good pressure, decreasing pressure, low pressure, increasing pressure and finally, good pressure will be encountered. If starting the shower when the pressure is on the low side, the watering can effect will be present until the pressure drops to its lowest setting and then gradually returns to normal. Surely not a nice way to start one's day.

Figure 33-A shows the typical one pressure switch arrangement that most simple systems employ. Figure 33-B shows the upgraded system consisting of an added flow switch connected in series with a high pressure switch. The flow switch and the high pressure switch are then connected in parallel with the standard high/low pressure regulating system. These in turn are connected in series to a high pressure limit switch. A high limit switch must always be installed on any system to prevent the possible explosion of the pressure tank and associated piping systems. With this arrangement, the pressure pump kicks in whenever any faucet or tap is opened. This assures good pressure throughout the plumbing system whenever the flow switch detects water usage. The settings on the new high pressure control switch that is connected in series with the flow switch are set more closely together than that of the original pressure switch. The high limit may be set to 90 psi while the low limit is set near 75 to 80 psi. This type of set up should ensure that the pressure pump does not cycle on and off unnecessarily during water usage.

WELL SYSTEM PRESSURE PUMP CONTROL

33-A

NORMAL SYSTEM

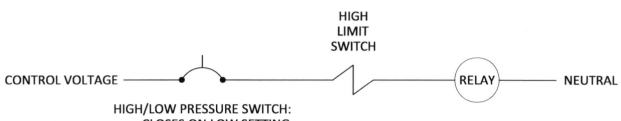

HIGH
LIMIT
SWITCH

CONTROL VOLTAGE ——————⌒—————————⚡—————(RELAY)——— NEUTRAL

HIGH/LOW PRESSURE SWITCH:
- CLOSES ON LOW SETTING
- OPENS ON HIGH SETTING

33-B

IMPROVED SYSTEM

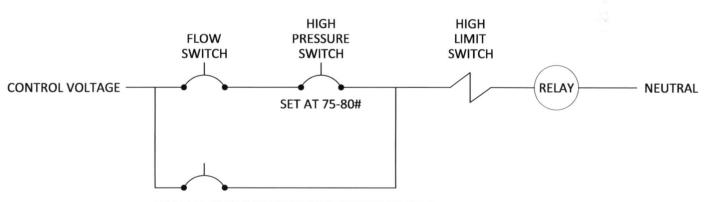

FLOW
SWITCH

HIGH
PRESSURE
SWITCH

HIGH
LIMIT
SWITCH

CONTROL VOLTAGE

SET AT 75-80#

RELAY

NEUTRAL

NORMAL HIGH/LOW PRESSURE SWITCH SETTING:
- 40# LOW
- 80# HIGH

OTHER PROPRIETARY SYSTEMS

There are many other proprietary systems out there on the market today. More and more manufacturers and investors feel that the now is the time to pursue this technology. With the popularity of tablets, laptops and mobile devices being able to communicate with these devices, the time seems ripe for technological advance in this area. The existing smart house ®[7] systems currently available, use proprietary high tech devices to control lighting, appliances and space conditioning through PC based software on your desktop, laptop, tablet or mobile device. These systems can all get rather expensive really quickly, so consequently not all builders offer these services. The other downside to some proprietary systems is called functional obsolescence. What may be popular and fashionable today may not be so in ten years time. Some of these companies may lack adequate funding and resources to support the older products or simply may no longer be in business. Trying to find a part or component that hasn't been manufactured in ten years may turn out to be a daunting task, maybe even difficult or impossible.

With hardwired techniques using standard cabling and economically priced wiring devices, the infrastructure will already be in place. The basic devices needed to accomplish the desired results will still be around 10, 20 or 30 years down the line. If something is left out now, it can always be easily addressed later using high tech gadgets on a limited basis.

X-10 TECHNOLOGY

X-10 ®[8] technology has been around for well over thirty-five years. This technology works by transmitting coded signals over the existing power lines and is known as power line carrier technology. Specialty coded switches and wiring devices are available that can do a multitude of both simple and complex tasks. When hardwiring is not practical, X-10 technology is certainly an option. These wiring devices can range from the inexpensive X-10 style and other economical brands licensed through X-10 and marketed under the names of Radio Shack ®[9] and Stanley ®[10] or higher priced Leviton ®[11]. The Leviton devices are somewhat more fashionable looking and as such come with a fancier price tag. There are also other high end manufacturers that cater to custom home wiring but their devices are priced accordingly.

X-10 technology is great but in my opinion, does have it's limitations and problems. The transmitted coded signals need to be of a sufficient magnitude when they reach the designated device. Weak signals due to excessive wire lengths in larger homes can become problematic. There are bridges and amplifiers available to help with these difficulties but noise can be another much more serious concern. Noise on the power line interferes with the transmission of coded signals and is more troublesome and harder to overcome. Sometimes the amplifiers needed to boost the signals also boost the noise levels. In addition, if a home is supplied by a transformer that also supplies other surrounding homes, that system as well as others on the same local grid can be affected, especially if the other homes use these same types of controls. There are filters that can be placed on the incoming power lines but again, this can get rather expensive.

---

7   Smart House is a registered trademark of Duva, Bruno J
8   X-10 is a registered trademark of X-10 Ltd
9   Radio Shack is a registered trademark of TRS Quality, Inc.
10 Stanley is a registered trademark of Stanley Logistics, L.L.C.
11 Leviton is a registered trademark of Leviton Manufacturing Co., Inc.

COMMON ELECTRICAL SYMBOLS & DIAGRAMS

⊖	DUPLEX RECEPTACLE, 120 VOLT		✕	CONNECTION POINT
⊕	QUADRAPLEX RECEPTACLE, 120 VOLTS		—Ⓜ—	COIL
⊘	240 VOLT RECEPTACLE			AC SOURCE
S	SINGLE-POLE SWITCH			TRANSFORMER
	SINGLE-POLE SWITCH			GROUND
SS	2-POLE SWITCH		⊕	LIGHT OUTLET
	2-POLE SWITCH		⊕	RECESSED LIGHT
SSS	3- POLE SWITCH		ⓈⒹ	SMOKE DETECTOR
	3- POLE SWITCH		△	TELEPHONE OUTLET
			▲	DATA OUTLET
S₃	3-WAY SWITCH			BATTERY
	3-WAY SWITCH		—∿∿—	RESISTOR
S₄	4-WAY SWITCH			VARIABLE RESISTOR
	4-WAY SWITCH (D.C. SWITCH)			POTENTIOMETER
	CIRCUIT BREAKER, SINGLE POLE		—‖—	CAPACITOR
	CIRCUIT BREAKER, TWO POLE			VARIABLE CAPACITOR
	FUSE		Ω	OHM
	NORMALLY OPEN PUSH BUTTON SWITCH			
	NORMALLY CLOSED PUSH BUTTON SWITCH			
	NORMALLY OPEN CONTACT			
	NORMALLY CLOSED CONTACT			

GLOSSARY

2-conductor cable	An outer jacketed cable containing 2 individual insulated wires within
2-pole switch	A switch having 2 internal switching contacts
3-conductor cable	An outer jacketed cable containing 3 individual insulated wires within
3-way switch	A switch used to control lighting or power from 2 locations
3-position switch	A switch having 3 positions, up, down and center off
3-phase loads	An electrical apparatus requiring 3 power conductors, 120° apart
3-pole switch	A switch having 3 internal switching contacts
3-phase power	Electricity supplied from 3 power conductors, 120° apart
4-way switch	A switch used to control lighting or power from 3 or more locations
ac	Alternating current (AC)
ac "OR" gate	Used to supply AC output whenever either 2 input conductors have power
ac electrolytic capacitor	An electrical component that resists any change in ac voltage
armored cable	An electrical cable having an outer jacket of steel or aluminum
bladder tank	A tank having a rubber style internal bladder to control water pressure

BX	**A type of armored electrical cable**
back fed	**Feeding power back into the power source**
center position off	**A 3 position toggle switch having its center position off**
common terminal	**An electrical point that is common to all other points on the device**
cable jacket	**The outer cover or jacket of an insulated cable assembly**
coil	**The part of an electrical contactor which when energized actuates the device**
current	**Also referred to as amperage flowing in a conductor**
cable clamps	**A fitting used to attach a cable to an electrical box**
California style 3-way	**A non code complying manner used to control lighting from 2 locations**
dc	**Direct current (DC)**
dc" OR" gate	**A logic device creating an output when either 2 inputs are present**
dead ended	**When wires in the cable are terminated on the switch with no splices**
double-pole-double throw switch	**A switch having 6 terminals, 2 center input terminals and 4 output terminals**
duplex receptacle	**Two outlet devices on one yoke or strap**
Edison base plug fuse	**A screw in type plug fuse rated up to 30 amps**
electrical code	**The National Electrical Code**
flow switch	**A switch that actuates upon detecting liquid flow in a pipe**

functional obsolescence	No longer considered technologically current, obsolete
ground fault circuit interrupter	An electrical device intended to interrupt the circuit whenever a pre-determined current flow to ground is detected
high pressure switch	A switch contact that opens or closes when high pressure is detected
high/low pressure regulating system	A combination of a low pressure switch and a high pressure switch that regulates the water pressure in a well supplied water system
high pressure limit switch	A safety switch that is designed to open upon sensing high pressure
hot conductor	A wire that is energized by voltage
"ice cube" style relay and socket	A clear plastic relay that is designed to control certain electrical functions that plugs into a mating socket where connections are made
internal jumper wires	Wires inside a switch that connect internal points on the switch
inverted	Turned upside down 180°
knob and tube	Old style residential wiring using porcelain knobs and tubes
light outlet box	An electrical outlet box designed to attach a light fixture
lineman's pliers	A style of pliers designed to cut and twist electrical wires
magnitude	The extent of destruction
magic triangle	A divided triangle shaped to calculate 3 factor formulas
maintained contact	A switch that remains closed until intentionally opened
momentary contact	A switch that remains closed as long as the switch toggle is held

momentary switch	A switch incorporating momentary contacts
multi-pole switch	A switch capable of controlling 2 or more electrical loads
multi-wire branch circuit	A circuit where 2 or 3 phase conductors share the same neutral
neutral conductor	The intentionally grounded circuit conductor that's colored white
NMB cable	Non-metallic sheathed cable type B also known as ROMEX
Ohms Law	An electrical formula used to calculate voltage, amperage and resistance
override	To force a particular desired action or result
polarity	Positive or negative voltage associated with direct current (DC)
power line carrier technology	A system whereby coded signals are generated and transmitted over existing power lines to control coded electrical device receivers
pressure switch	A switch designed to open or close upon a preset pressure setting
psi	Abbreviation for pounds per square inch
pressure pump	A pump designed to raise pressure in a well supplied water system
power	Also known as wattage or power consumption
power formula	$P = I \times E$, power or wattage is the product of current multiplied by voltage
phase conductors	The energized wires providing electrical power, 120° or 180° apart
proprietary wiring devices	Licensed technological wiring devices and controls
pigtailed	Multiple spliced wires having only one connection tail

recessed lighting fixture	A light fixture with its main body entirely located inside the ceiling
relay	An electro-mechanical device used to control electrical functions
reversing switch	A switch used to reverse direction on a direct current (DC) motor
ROMEX	Non-metallic sheathed cable, also known as NMB
resistance	The property of a conductor that resists current flow
swapping	Reversing the wire landing sequence on a 3-way switch
single pole double throw switch	A single pole switch having 3 terminals, one center terminal in and 2 terminals out
standard single pole	A simple on/off switch used to control lighting
switch box	An electrical junction box housing a switch
switch leg	The conductor leaving the switch and feeding the light fixture
spliced	The mechanical and electrical joining of 2 or more conductors
screw shell	The outer threaded section of a lamp holder
single receptacle	One outlet wiring device on one yoke or strap
switch loop	The switches and conductors controlling the lighting
terminals	The attachment points on a wiring device
travelers	The two conductors between 3-way and 4-way switches
toggling	Changing the position of a switch from up to down or vise versa

transmitted coded signals	Used to control proprietary wiring devices on existing circuits
tolerance	The ohmic accuracy of a resistor expressed in percentage
toggle	To physically change the position of a switch
transformer	An electrical apparatus used to either step up or step down AC voltage
transfer switch	Used to change electrical input from utility power to generator power
type S fuse adaptor	A threaded insert used to limit fuse sizes in Edison base plug fuse sockets
unbalanced multi-wire branch circuit	A circuit served by two phase conductors sharing the same neutral and having different amperage on each phase conductor
voltage	Electrical pressure expressed in volts
voltage drop	The reduction in voltage/pressure caused by resistance in the conductor
volts	A quantity of electrical pressure such as 120 volts AC
wattage	The power consumed in an electrical circuit
wire twisting or pre-twisting	The physical twisting of individual conductors together prior to installing the wire nut or fastener
wire nut	An insulated electrical fastener used to splice conductors together
X-10	A proprietary system using coded signals transmitted over existing power lines to control coded receivers

Made in the USA
San Bernardino, CA
14 March 2018